MACMILLAN WORK OUT SER

Engineering Thermodynamics

The titles in this series

- Biochemistry
- Dynamics
- Electric Circuits
- Electromagnetic Fields
- Electronics
- Elements of Banking
- Engineering Materials
- Engineering Thermodynamics
- Fluid Mechanics
- Mathematics for Economists
- Operational Research
- Organic Chemistry
- Physical Chemistry
- Waves and Optics

MACMILLAN WORK OUT SERIES

Engineering Thermodynamics

G. Boxer

First published 1987
Reprinted 1990

Published by
MACMILLAN EDUCATION LTD
Houndmills, Basingstoke, Hampshire RG21 2XS
and London
Companies and representatives
throughout the world

Typeset by TecSet Ltd,
Wallington, Surrey
Printed in Hong Kong

British Library Cataloguing in Publication Data
Boxer, G.
Work out engineering thermodynamics.—
(Macmillan work out series)
1. Thermodynamics — Problems, exercises,
etc.
I. Title
536′.7′02462 TJ265
ISBN 0–333–43664–4

Contents

Preface

This is not a textbook in thermodynamics; it is a revision tutorial volume. It is assumed that the student has completed a course of lectures in the first year and is reasonably familiar with all the terms in use here. The student should then be able to use this volume for revision purposes in preparing for an examination on the first year's work.

Consequently, no derivation of fundamental equations (e.g. the steady flow energy equation) will be found within these pages. The student is referred to recommended texts given at the end of this volume for this fundamental work or to notes taken during lectures. One exception is in momentum; a reasonably full treatment is given in Chapter 6, making use of the concept of free body diagrams, to help the student to a clearer understanding of a topic that is generally difficult to assimilate.

I have done all I can to make the technical language clear because the correct use of physical symbols (and the dimensions associated with them) is essential to a proper understanding of the subject. I stress within these pages the invaluable help that may be obtained in any fundamental reasoning with the use of unity brackets, which I cannot commend too highly.

I have included a standard procedure for the solution of all problems in thermodynamics wherein the student is encouraged to answer certain leading questions, the correct answers to which will reveal the uniformity of approach possible in this subject. This assumes that the main work of lecturing is over and that the student is now faced with the re-examination, assimilation and application of this work in a logical way. I really believe the standard procedure will be of assistance here.

1987 G. B.

Acknowledgements

I acknowledge with gratitude the permission of Messrs Basil Blackwell of Oxford to use extracts from tables of properties by Rogers and Mayhew in the solutions presented in this book.

I also acknowledge with gratitude the help of my colleagues in the department of mechanical engineering at the University of Aston in Birmingham, particularly those who remain to teach this demanding subject.

Nomenclature

A	area
A_S	surface area
a	dimension
B	dimension
b	dimension
C	heat capacity
c	specific heat capacity *or* dimension
c_p, c_v	the principal specific heat capacities (isobaric, isochoric)
c.s.	control surface
D	diameter
d	prefix for differential quantity (e.g. dx)
E	internal energy (preferred alternative to U in British Standard)
e	specific internal energy ($e = E/m$)
F	force
f	symbol denoting liquid line (from German *Flussigkeit*)
g	gravitational acceleration
g	symbol denoting saturated vapour line
H	enthalpy
h	specific enthalpy ($h = H/m$) *or* local coefficient of heat transfer
i	constituent suffix in gas mixtures
K	kelvin unit of temperature on the absolute scale
k	thermal conductivity
L	length
l	length
m	mass
m_v	relative molar mass
$\dot{m}$	mass flow rate ($\dot{m} = \mathrm{d}m/\mathrm{d}t$)
N	speed of rotation
n	polytropic index *or* number of kmols of a substance
P	brake load on an engine
p	pressure
Q	heat transfer
$\dot{Q}$	rate of heat transfer ($\dot{Q} = \mathrm{d}Q/\mathrm{d}t$)
$\dot{Q}'$	rate of heat transfer per unit length
$\dot{Q}''$	rate of heat transfer per unit area
$\dot{Q}'''$	rate of heat transfer per unit volume
q	specific heat transfer ($q = Q/m$)
R	spring rate or specific gas constant or thermal resistance
R_0	universal gas constant
r	radius
r_w	work ratio

S	entropy
s	specific entropy ($s = S/m$)
t	time
t_0	time constant in heat transfer in unsteady state
T	temperature
U	overall coefficient of heat transfer
u	velocity
V	volume
$\dot{V}$	volume flow rate ($\dot{V} = \mathrm{d}V/\mathrm{d}t$)
v	specific volume ($v = V/m$)
W	work transfer
$\dot{W}$	rate of work transfer or power ($\dot{W} = \mathrm{d}W/\mathrm{d}t$)
w	specific work transfer ($w = W/m$)
x	principal coordinate *or* quality (dryness) of vapour *or* dimension
y	dimension
z	datum height *or* dimension
α	angle
β	coefficient of cubical expansion
γ	ratio of c_p/c_v
δ	prefix for incremental change (e.g. δx)
η	efficiency
θ	angle *or* temperature difference
μ	absolute or dynamic viscosity
π	as in area of circle = πr^2
ρ	density or resistivity
σ	Stefan–Boltzmann constant
Σ	sum of
Σ	sum of round a cycle
ϕ	angle
ω	angular velocity

Dimensionless groups

Nu	Nusselt number ($= hL/k$)
Pr	Prandtl number ($= c_p \mu/k$)
Re	Reynolds number ($= \rho u L/\mu$)
Gr	Grashof number ($= L\beta g\theta u^2/\mu^2$)
Pe	Peclet number ($= uL/\alpha$ where $\alpha = k/\rho c$)
Ec	Eckert number ($= u^2/c\theta$)

Other symbols

$\equiv$	equivalent to
$\propto$	proportional to
$>$	greater than
$<$	less than
$\geqslant$	greater than or equal to
$\leqslant$	less than or equal to
$\uparrow$	increasing
$\downarrow$	decreasing
∞	infinity
$\simeq$	approximately equal to
$\hat{}$	maximum (e.g. $\hat{p}$ = maximum pressure)
$\emptyset$	a function of

1 Introduction

1.1 Definitions

A clear grasp of thermodynamics fundamentals does not need a knowledge of advanced mathematical techniques. It is, however, essential to have a very clear understanding of the terms used and the manner in which the first and second laws of thermodynamics can be applied.

We begin, then, with a series of definitions designed to give the student a rigorous basis for the work that follows.

System

A quantity of matter (or mass) in equilibrium whose behaviour is under investigation. (Note that the word equilibrium is defined further on.) For ease of analysis a closed system is enclosed by a control volume or surface (also defined below).

An example of a closed system is a mass of air and petrol vapour, intimately mixed, which is being compressed in an internal-combustion engine.

In steady flow the contents of a given region in space through which mass is passing are an example of an open system.

Fluid

A substance or mixture of substances in the liquid or gaseous state.

Property

An observable or calculable characteristic of the system, e.g. pressure p, tempera- T, velocity u, internal energy E. Note that pressure and temperature are observable (can be measured) and internal energy has to be calculated. Note also that pressure and temperature are intensive properties (do not depend on mass) but internal energy is not intensive. This leads to the definition of intensive and extensive properties.

Intensive properties are properties per unit mass in the case of volume, internal energy, enthalpy and entropy. They are always written in lower-case letters, although the latter do not always imply specific properties since pressure is written in lower case as well. Extensive properties are given in upper-case letters. Note that extensive properties are additive but intensive properties are not.

Furthermore, velocity and datum height are mechanical properties and viscosity and thermal conductivity are transport properties.

State

The thermodynamic state of a system is defined completely by the knowledge of two independent and intensive properties, e.g. pressure and specific volume or temperature and specific internal energy. Note that pressure and temperature do not necessarily define the state since they are not independent during the evaporation of a liquid.

Process

A process is a change in the state of a system. Two kinds of process are important in this study, as follows:

(a) *Non-flow process* A fixed identifiable mass of fluid undergoes a change of state with no mass transfer across the control surface. (For example the compression of the mixture in an internal-combustion engine.)
(b) *Steady-flow process* A uniform mass flow rate from one state to another. Wherever the control surface is drawn for examination mass must cross at inlet and outlet. (For example steady compression of air in a jet engine.)

Control Surface (Volume)

The control surface encloses a region in space (the control volume) as follows:

(a) Inside the space there is a fixed mass of fluid whose behaviour is under investigation (as in a non-flow process).
(b) Across the space there may be a mass transfer at inlet and outlet whose thermodynamic state is changing and is under investigation as for a flow process (steady or unsteady).

Examples are given in the diagrams. Note that in the non-flow process in the engine cylinder the control volume changes its shape as the piston moves, but the mass is fixed for compression (or expansion). However, in the turbine the control volume is fixed in space and size, and the mass flow crosses at 1 and 2.

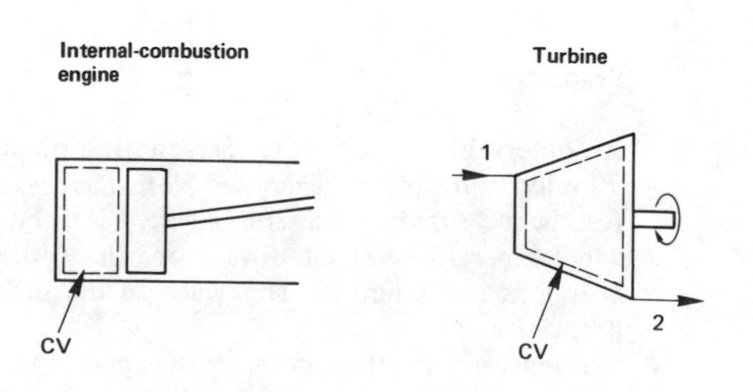

Surroundings

The rest of the universe outside the control volume.

Cycle

This is a process with identical initial and final states. Examples are: a steam power plant (see diagram) consisting of a steam generator (G), followed by a turbine (T), condenser (C) and feed pump (P) and back to G; and a closed-circuit gas turbine consisting of a compressor (CM) followed in order by a heater (H), a turbine (T), a cooler (C) and back to the compressor.

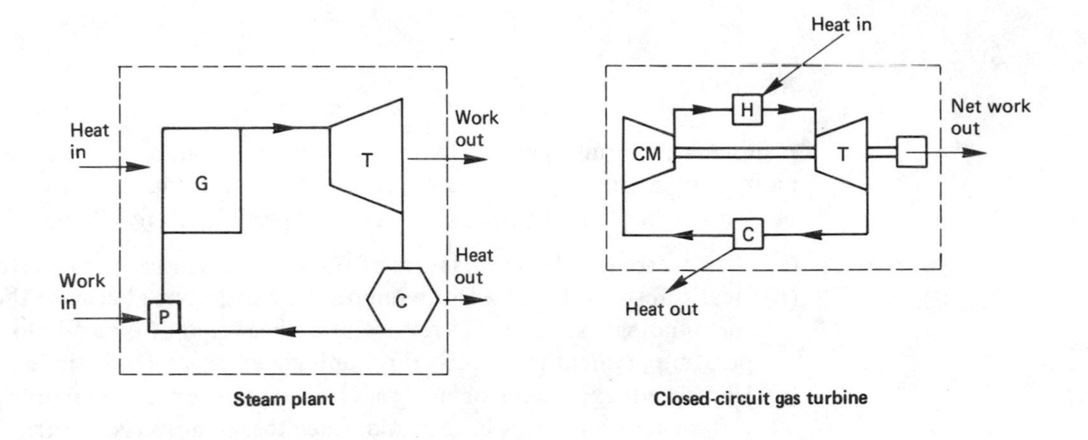

Steam plant

Closed-circuit gas turbine

Equilibrium

A fluid is in equilibrium when no energy transfers occur between different parts of the mass, if the latter is isolated from its surroundings; i.e., the thermodynamic properties must be constant. Three kinds of equilibrium need identification, as follows:

(a) *Thermal equilibrium:* requires a uniform temperature (otherwise there will be a heat transfer from the higher-temperature part to the lower-temperature part).
(b) *Mechanical equilibrium:* requires a uniform pressure (otherwise there will be a mass transfer from the high-pressure part to the lower-pressure part).
(c) *Chemical equilibrium:* requires a uniform chemical composition (otherwise there will be a chemical reaction or a change in the chemical composition in the system).

Thermodynamic equilibrium exists when *all three above* hold good.

Interaction

This is any event occurring at the control surface which causes a change in state of the system. The two prime examples are heat and work transfer. Such an event is depicted in the accompanying diagram and the internal energy changes.

The diagram depicts a non-flow process with addition of heat and work transfer. E_1 rises to E_2.

Phase

For example, ice, water and steam are respectively the solid, liquid and vapour phases of H_2O.

Note that 'gas' has a special meaning in this context, and refers to the special case of a vapour well above the fluid's critical temperature and not to all the vapour states – see Chapter 4.

Reversible

A process is thermodynamically reversible when it can be made to take the reverse path through the same series of measurable equilibrium states as it took on its outward path. Thus a process cannot be reversible if any of the following occur:

(a) Viscous friction between layers of fluid which will cause irreversibility.
(b) Heat transfer through a finite temperature difference between the fluid and its surroundings, since in the reverse sense heat cannot be made to 'climb' a temperature gradient unaided – this implying external irreversibility.
(c) Unresisted expansion or any rapid expansion or compression giving rise to turbulence or vortices in the fluid, since these imply viscous friction.
(d) Combustion or chemical reaction with a finite temperature difference between the fluid and its surroundings.

Note that (a) and (c) constitute mechanical irreversibility, (b) constitutes thermal irreversibility and (d) consistitutes chemical irreversibility.

Clearly, all real processes are irreversible, but ideal reversible processes can be imagined with which real processes can be compared.

One example is the expansion in a high-speed turbine which can be imagined to take place ideally both reversibly and without heat transfer (this implying a constant value of entropy as shown in Chapter 9).

Real expansion in such a turbine leads to a gain in the fluid entropy, and the ideal and real expansions are compared using an efficiency of isentropic expansion. The use of the word isentropic leads to the next set of definitions.

Adiabatic

This means that no heat transfer takes place between the system and its surroundings (correspondingly, diabatic means the opposite – i.e. the heat transfer is finite).

Isothermal

The system temperature is constant.

Isobaric

The system pressure is constant.

Isochoric

The system specific volume is constant.

Isentropic

The system specific entropy is constant.

Finally, it may be helpful to point out that although reference has been only to the first two laws of thermodynamics there is a third law which is treated in main texts and should be assimilated but is of detailed value in the later treatment of the subject and of less significance in the first year.

It should, perhaps, also be pointed out that the subject of thermodynamics is really a study of the effects of heat and work upon the system properties, or, conversely, the effects of a change in the system properties upon the heat and work transfers.

1.2 Problem Solution Approach

Given the preceding definitions, the following procedure is suggested to help the student tackle any problem in thermodynamics in a consistent and logical manner. I should point out at once that this method is used only in Chapters 13 and 14 because the subject matter is divided up into readily identifiable packets all appropriately labelled, e.g. non-flow processes, flow processes, etc., and the questions suggested below are already answered for the student.

In an examination paper (as in Chapter 13 or 14) such a procedure would be of considerable value since no labels are given there and selecting these and the consequent treatment is a matter you will have to decide upon *before* doing anything else. Such a method absolves you from splitting up the work into artificial packets with the need for you to attempt to 'remember' given formulae for each packet.

Equilibrium thermodynamics is *not* a compound of formulae. It has its basis in the application of three fundamental principles, as follows:

(a) The principle of conservation of mass.
(b) The principle of conservation of energy.
(c) The principle of conservation of momentum.

Here are the questions.

(a) What Kind of *Process*?

The choice lies between non-flow, steady flow and unsteady flow. The aim is to draw a control surface in the appropriate place and to decide if *mass* crosses this surface. If not then we are dealing with a non-flow process. If there is a steady mass flow then we have a steady-flow process and if the mass flow is unsteady (as in the emptying of a cylinder through an orifice) then we have an unsteady-flow process.

(b) What Kind of *Fluid*?

The choice lies between subcooled liquid, saturated liquid, partly evaporated liquid (or wet vapour), saturated vapour, superheated vapour, and gas.

(c) Have you Drawn a *Sketch* of Events?

It is essential to sketch a field of state (e.g. pressure versus specific volume) showing the relevant fluid boundaries and marking in the appropriate states.

(d) Do You Need the *Mass Continuity* Equation?

This is $$\dot{V} = \dot{m}v = uA,$$

where the symbols have the meaning listed in the nomenclature (pp. ix–x). (Refer to the standard texts for derivation.)

(e) Do You Need the *Energy* Equation?

This is, in effect, using the first law of thermodynamics. Thus for a non-flow process from state 1 to state 2 for unit mass of fluid in the system,

$$_1q_2 - {}_1w_2 = e_2 - e_1.$$

In steady flow

$$_1\dot{Q}_2 - {}_1\dot{W}_2 = \dot{m}\,[(h_2 - h_1) + \tfrac{1}{2}(u_2^2 - u_1^2) + g\,(z_2 - z_1)]$$

which in reduced form becomes

$$_1\dot{Q}_2 - {}_1\dot{W}_2 = \dot{m}\,(h_2 - h_1).$$

when kinetic and potential energy terms are negligible. (Refer to standard texts for derivation.)

(f) Do You Need to Use the *Momentum* Equation?

In this instance I have seen fit to give a reasonably full treatment suitable for a first-year course in Chapter 6, using the concept of the free-body diagram to assist you in gaining a clearer picture of this equation, which is often overlooked and is very often dimly understood. The principle deals with the calculation of *forces*.

(g) Have You Used the Correct *Language*?

For example, w for specific work transfer (in kJ/kg),
W for work transfer (in kJ/kg),
$\dot{W} = \dfrac{dW}{dt}$ = rate of doing work (or power), given in kW since the dimensions of a differential coefficient are determined by drawing a line between the differential operators and the physical quantities they qualify (e.g. here $\dfrac{d:W}{d:t}$ gives $\dfrac{W}{t}$ or kW).

(h) Have You Ensured Correct *Dimensional Reasoning*?

This is linked very closely with (g) above, since correct language leads to correct dimensions for the careful student. The following example serves to illustrate the point.

From the steady-flow energy equation (potential energy ignored),

$$ {}_1\dot{Q}_2 - {}_1\dot{W}_2 = \dot{m}\ [(h_2 - h_1) + \tfrac{1}{2}(u_2^2 - u_1^2)]. $$

Suppose we wish to calculate u_1 for a fluid which is being decelerated to rest in a nozzle with no heat or work transfer.

Then $u_1 = \sqrt{2\ (h_2 - h_1)}$ from the above after cancellation, and if h is in kJ/kg then u will, in the first instance, be in $\sqrt{\text{kJ/kg}}$ and not in the required units of m/s.

With the use of dimensional reasoning and *unity brackets* we can easily transform the above calculation to yield m/s. Thus Newton's law states that

$$ 1\ \text{N} = 1\ \text{kg} \times 1\ \frac{\text{m}}{\text{s}^2} $$

or

$$ \left[\frac{\text{N s}^2}{\text{kg m}}\right] = [\text{UNITY}] = \left[\frac{\text{kg m}}{\text{N s}^2}\right] $$

and anything multiplied by unity (or its reciprocal) is unchanged.

Furthermore, $1\ \text{kJ} = 1\ \text{kN m}$

or

$$ \left[\frac{\text{kJ}}{\text{kN m}}\right] = [\text{UNITY}] = \left[\frac{\text{kN m}}{\text{kJ}}\right]. $$

Thus if we now rewrite the above with two unity brackets as follows we get

$$ u_1 = \sqrt{\underbrace{2\ (\text{a number})\ \frac{\text{kJ}}{\text{kg}}}_{(h_1 - h_2)}\ \underbrace{\left[\frac{\text{kN m}}{\text{kJ}}\right]}_{\text{UNITY}}\ \underbrace{\left[\frac{\text{kg m}}{\text{N s}^2}\right]}_{\text{UNITY}}} $$

or

$$ u_1 = \sqrt{2\ (\text{a number}) \times \text{k}\ \frac{\text{m}^2}{\text{s}^2}} = \sqrt{2000\ (\text{a number})}\ \frac{\text{m}}{\text{s}}. $$

Notes

(a) We have chosen those unity brackets which will give the required final units (i.e. m/s). Any other substitution would fail in this regard.

(b) The figure of 1000 has appeared in the result due to the cancellation of dimensions and the non-thinking student, not listing the dimensions, will arrive at a result $\sqrt{1000}$ i.e. 31.62 times too small!

Consider now a second example in the calculation of T from the use of the characteristic gas equation, namely

$$ T = \frac{pV}{mR}, $$

with $p = A$ bar
$V = B\ \text{m}^3$
$m = C$ kg
$R = D$ kJ/(kg K)

where A, B, C and D are pure numbers.

$$ T = \frac{A\ \text{bar} \times B\ \text{m}^3}{C\ \text{kg} \times D\ \dfrac{\text{kJ}}{\text{kg K}}} $$ (which does not yield K as required).

However 1 bar = $10^5 \dfrac{\text{N}}{\text{m}^2}$

and 1 kJ = 1 kN m (as before).

$$\text{Thus} \quad \left[\frac{10^5 \dfrac{\text{N}}{\text{m}^2}}{\text{bar}}\right] = [\text{UNITY}]$$

$$\text{and} \quad \left[\frac{\text{kJ}}{\text{kN m}}\right] = [\text{UNITY}]$$

$$\text{and } T = \frac{A\ \cancel{\text{bar}} \times B\ \cancel{\text{m}^3}}{C\ \cancel{\text{kg}} \times D\ \dfrac{\cancel{\text{kJ}}}{\cancel{\text{kg}}\ \text{K}}} \times \left[\frac{10^5 \dfrac{\cancel{\text{N}}}{\cancel{\text{m}^2}}}{\cancel{\text{bar}}}\right] \left[\frac{\cancel{\text{kJ}}}{\text{k}\cancel{\text{N}}\ \cancel{\text{m}}}\right] = \frac{AB}{CD} \times \frac{10^5}{10^3} = 100 \frac{AB}{CD} \text{ K.}$$

Note: This regularly occurring calculation is probably best carried out by changing the non-SI unit, bar, to an SI unit of pressure, kN/m^2, by multiplying by 10^2 to start with, and this is the method adopted in the book. The bar may not be an accepted SI unit but it is in widespread use and you must learn to handle it, as this is still accepted practice in the engineering professions.

This method of checking dimensions as you go will not resolve all your difficulties, but if my experience at Aston University over some thirty years is anything to judge by it will resolve most of them. Correct language and dimensions cannot be too strongly recommended.

The above procedure can be applied throughout an examination paper either wholly or in part for all questions in thermodynamics and two specimen papers are given with the application of this approach in Chapters 13 and 14. Note that heat transfer questions do not lend themselves to this approach since heat transfer has its unique laws of radiation, conduction and convection.

2 Systems

2.1 Introduction

In the following problems the standard engineering convention of signs for heat and work transfers is used, namely that heat is positive when its sense is from the surroundings to the system (see previous definitions) and negative when in the opposite sense. Work transfer is positive when the sense is from the system to the surroundings and negative in the opposite sense.

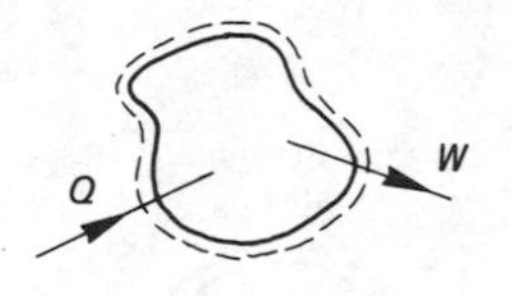

The distinction must be made between sense and direction in that both heat and work are scalar quantities, i.e. having magnitude and sense but *all* directions, unlike, for example, force which is a vector quantity having magnitude, *one* direction and sense.

2.2 Worked Examples

Example 2.1

Give the sense of q and w for one kilogram of air which flows so rapidly from a charged reservoir into a previously evacuated bottle that heat transfer may be neglected.

Answer

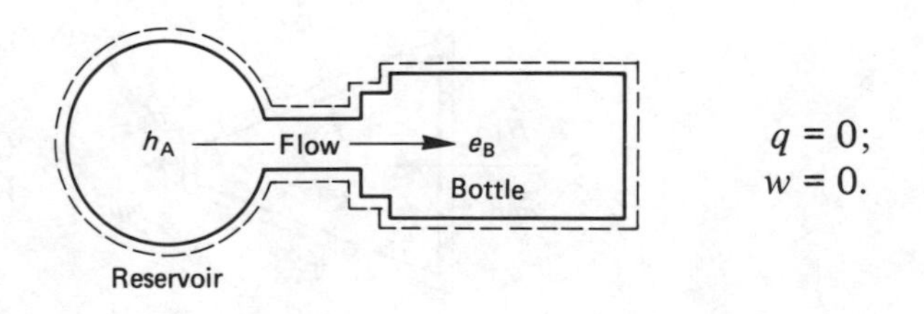

$q = 0;$
$w = 0.$

This example has particular significance in that it affords an opportunity to distinguish between the internal energy of a fluid which is all that remains at the end of this process in the bottle and the *enthalpy* of the air in the charging line outside.

Note that not only has the air outside got its own internal energy but it has also to possess enough extra energy to enable it to flow into the evacuated bottle. The outside air is assumed to flow from an infinite reservoir.

This extra energy which is numerically equal to the product of pressure and specific volume for the air (see standard texts for proof) is required to *push* the air into the bottle, i.e. *to cause a mass transfer*.

Enthalpy is defined as the sum of internal energy e_A (due to internal molecular vibration) and the product of pressure and specific volume $(pv)_A$ per unit mass, the latter term being written in whenever mass flow is involved.

Summarising, we may say here that

$$h_A = e_B \qquad \text{where } h_A = e_A + p_A v_A.$$

Example 2.2

Give the sense of Q and W for fluid contained within a rigid sphere into which a stirrer projects if the stirrer is rotated such that the internal energy of the fluid inside rises from E_1 to E_2. The surface of the sphere may be assumed to be a perfect insulator.

Answer

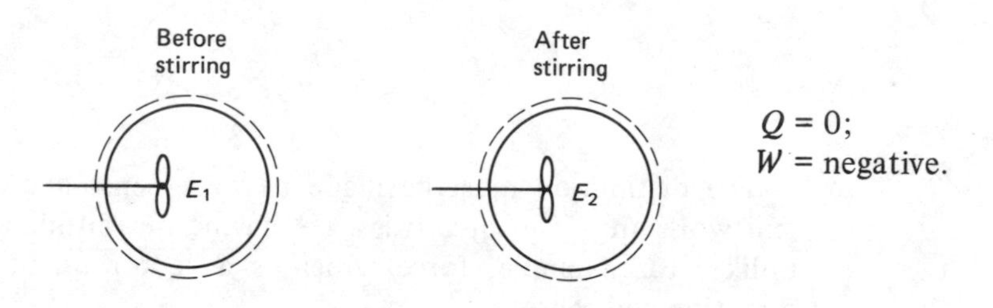

Note that work has transferred via the stirrer from the surroundings to the system (fluid inside sphere) and is thus negative by our convention.

Example 2.3

Give the sense of $\dot{Q}$ and $\dot{W}$ for a nozzle in which fluid flows from a high value of enthalpy to a lower value of enthalpy leading to the generation of a high-speed jet. Show all energy forms on your diagram.

Answer

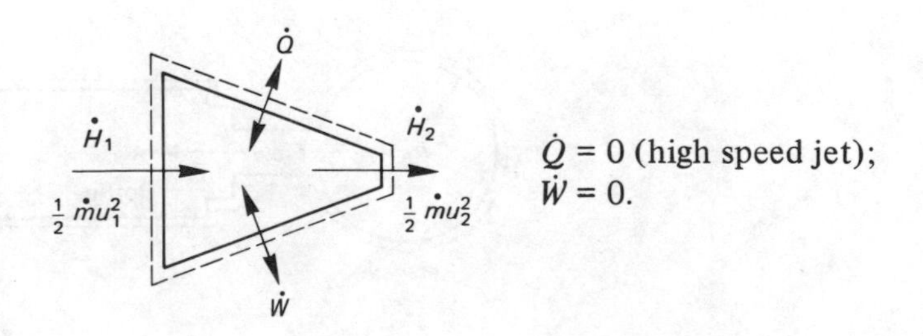

Example 2.4

Give the sense of Q and W for the contents of a bomb calorimeter in which there is a charge of both oxygen and fuel. A spark of very small energy (relative to combustion) causes the two reactants to combust and leaves products of combustion behind. The calorimeter may be assumed rigid and fully insulated.

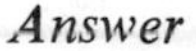

Answer

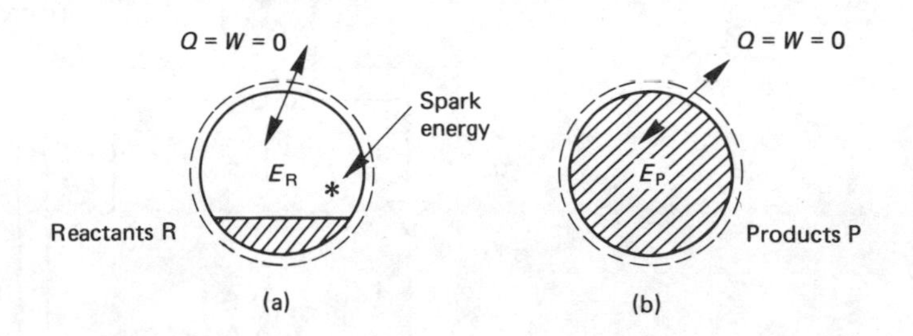

There is no heat or work transfer. However, we have to recognise that an energy release of considerable magnitude has taken place due to combustion.

What has actually happened is that the energy stored in the constituent atoms of the two reactants has been released when these two combine in combustion. No heat has crossed the system boundary (the surface of the calorimeter) but it is common practice to refer to the energy release in combustion here as the enthalpy of combustion. This is usual wherever energy is released into the fluid by virtue of hydrocarbon combustion with oxygen.

***Example** 2.5*

Diagram (a) shows schematically a simple vapour-compression refrigerator circuit together with the secondary circuits associated with the condenser and evaporator sections. Draw a system boundary or control surface and indicate on it all the energy transfers.

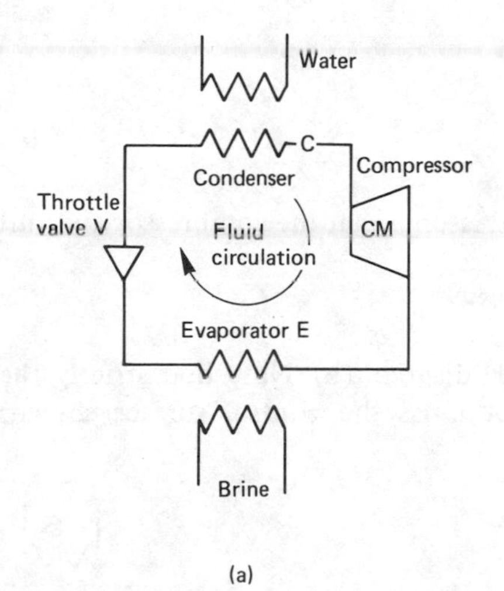

(a)

Answer

Diagram (b) shows the principal heat and work transfers associated with a simple refrigerator plant, neglecting extraneous energy transfer with the environment.

Note that the identical diagram can be drawn for a heat pump and, further, that in a refrigerator interest centres on the value of ${}_4\dot{Q}_1$ whereas in a heat pump it is on ${}_2\dot{Q}_3$.

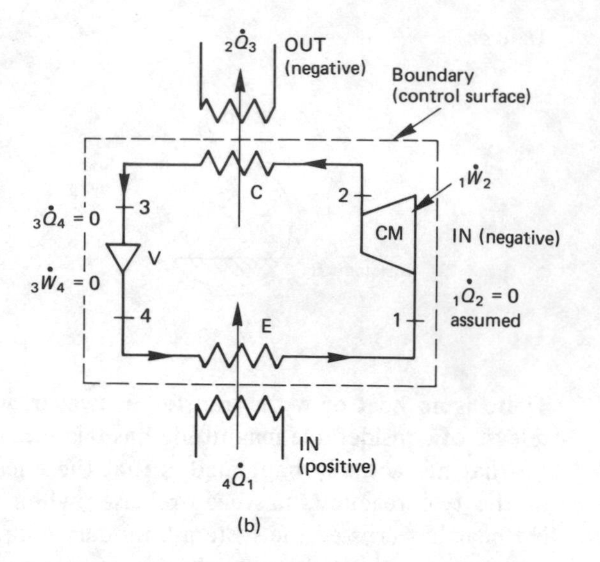

(b)

Example 2.6

Diagram (a) shows schematically a closed-circuit gas turbine plant. Draw an appropriate control surface and indicate on it the heat and work transfers.

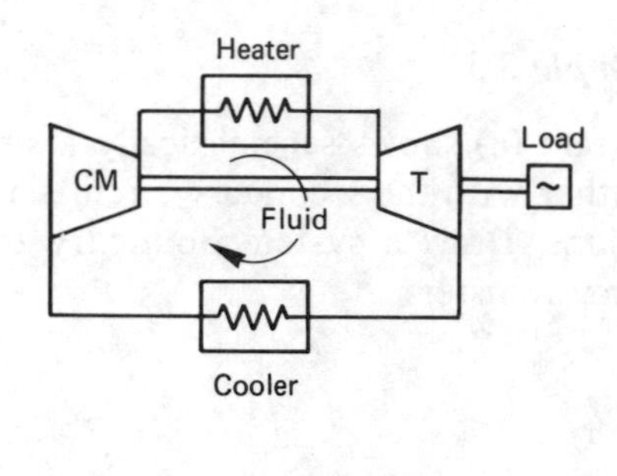

(a)

Assume that the compressor and turbine operate adiabatically.

Answer

See diagram (b). Note that strictly the work to drive the compressor section does not cross the control surface shown but would cross a control surface drawn

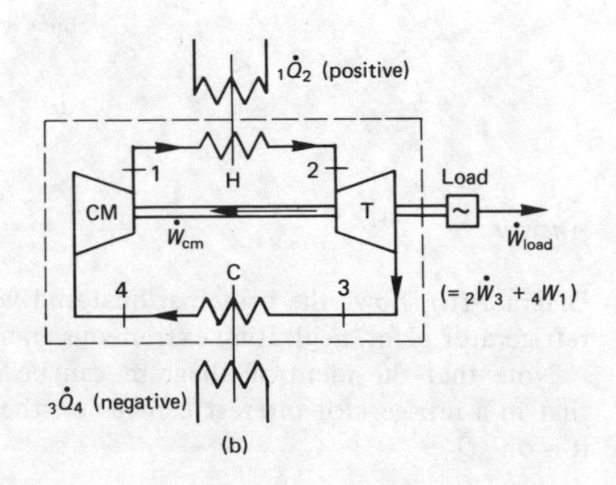

(b)

round the turbine alone. The turbine has to generate enough power to drive both the compressor section and the external load as well.

Example 2.7

A reheat steam power circuit consists of a steam generator, a superheater, an h.p. turbine, a reheater, an l.p. turbine, a condenser and a feed pump in that order.

Construct an appropriate schematic showing all components appropriately and with the use of a corresponding control surface mark on all the associated heat and work transfers in the circuit. Show any necessary secondary circuits. The turbines operate adiabatically.

Answer

See diagram.

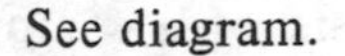

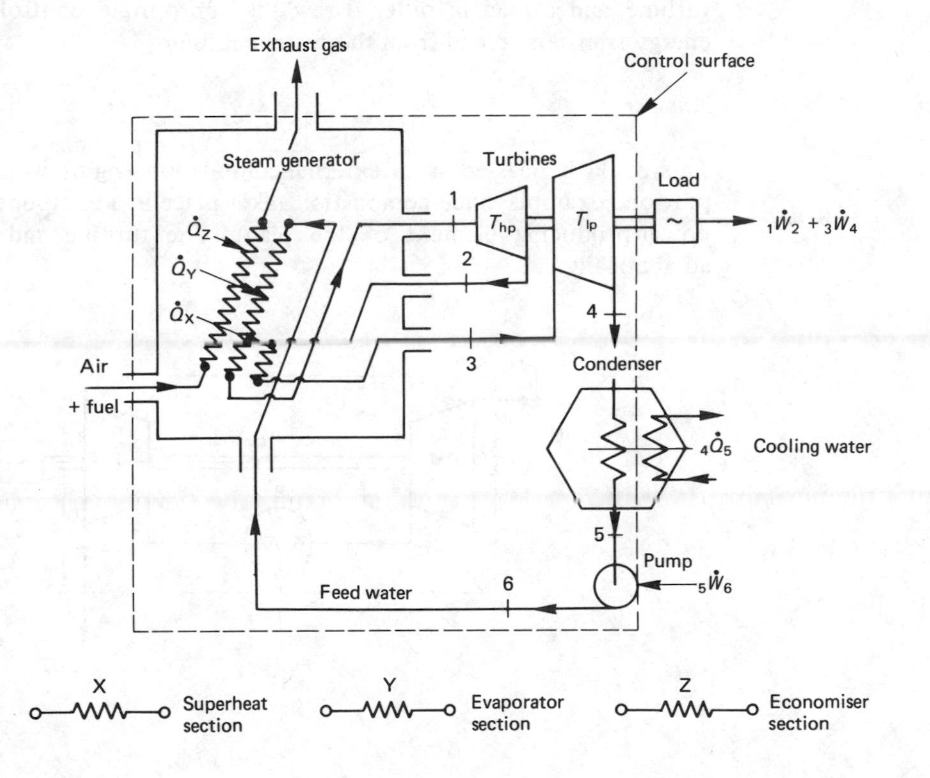

Example 2.8

Draw an appropriate schematic for an internal-combustion engine which burns a mixture of hydrocarbon fuel and air, rejects exhaust gas, delivers power to its crankshaft and rejects energy by heat transfer both to cooling water and also by radiation from its hot surfaces to the atmosphere. Mark on the diagram an appropriate control surface showing all heat, work and mass transfers.

Answer

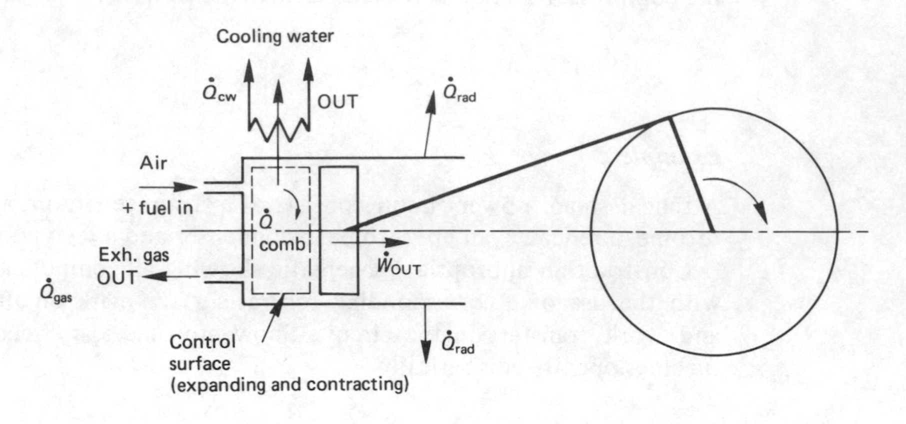

Example 2.9

A simple open-circuit gas turbine plant consists of a compressor, a burner, a turbine and a load in order. Draw the appropriate control surface showing all energy transfers to and from the system chosen.

Answer

This could be classed as an external-combustion engine when compared with the previous example since combustion takes place in a component separate from the power-producing element of the plant. The turbine and compressor operate adiabatically.

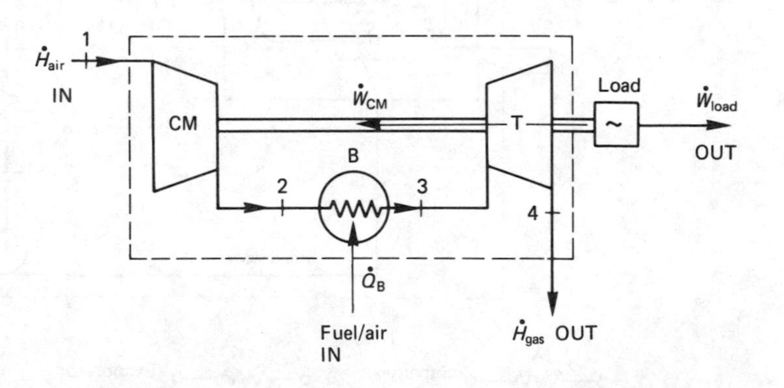

Example 2.10

A gas-cooled nuclear power reactor transfers heat through its core to carbon dioxide gas. The gas is pumped through channels surrounding the fuel elements in the core. This gas then passes out of the core to a heat exchanger and transfers energy by heat transfer to a water/steam circuit which is part of a steam power-producing circuit consisting of a turbine, a condenser and a feed pump.

Draw the appropriate control surface to embrace all salient energy transfers and mark these on the diagram paying careful attention to sense in each case. The turbine and feed pump operate adiabatically.

Answer

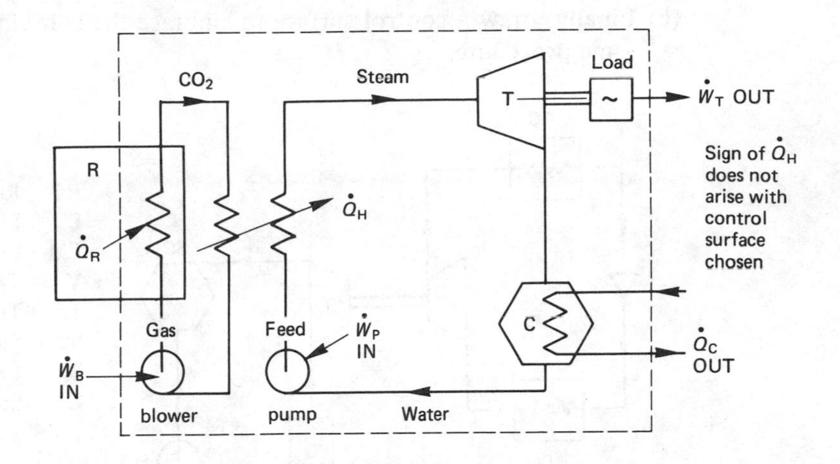

Exercises

1 Draw a sketch of a simple gas turbine jet engine layout consisting of a ram intake for initial compression, a main compressor section, a burner, a turbine and an expansion nozzle in that order. Draw on the sketch a control volume and indicate on your sketch all the relevant energy flows, assuming that there is no heat transfer from any component owing to the high-speed flow.

2 A two-stage feed-heating steam plant is shown diagrammatically in the diagram. Draw on this a suitable control surface and mark on all the relevant energy transfers, assuming that the turbines operate adiabatically.

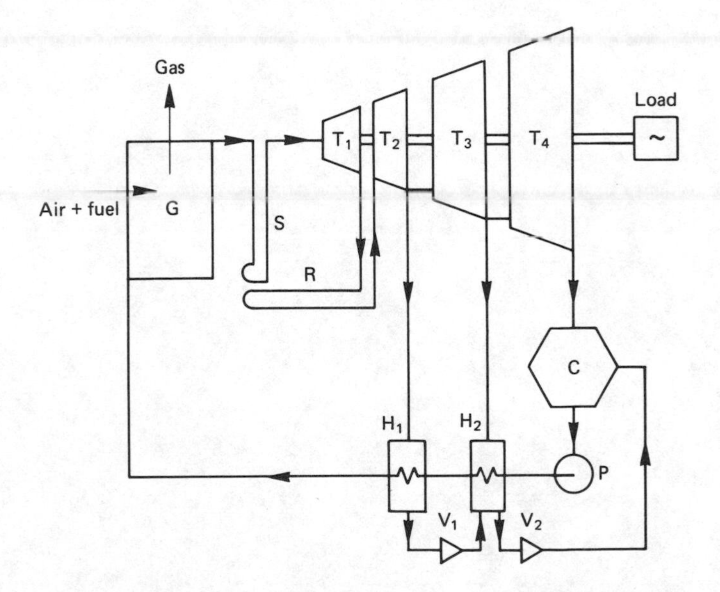

3 The diagram overleaf shows, schematically, a combined steam power/freon refrigeration plant coupled (T/CM) as given with no net work output.

(a) Draw a control surface around the refrigeration plant alone and mark on all the appropriate energy transfers, assuming that the compressor operates adiabatically.

(b) Draw a second control surface around the steam plant and proceed as in (a) above assuming now that T and P operate adiabatically.

(c) Finally, draw a control surface to embrace the combined plant and repeat the same procedure.

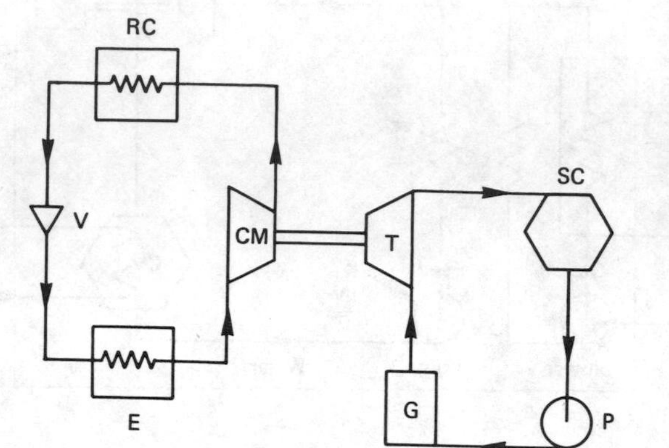

RC	Refrigeration condenser
E	Evaporator
CM	Compressor
V	Throttle
T	Turbine
SC	Steam condenser
P	Feed pump
G	Steam generator

3 Work Transfer

3.1 Introduction

Work is an energy transfer which occurs in a non-flow process when part of the system boundary suffers a displacement owing to the action of a pressure.

Correspondingly, work is done when a steady-flow, non-positive displacement device such as a turbine is rotated by virtue of a pressure gradient across the machine. In this case one can imagine the shaft of the turbine having a cord wrapped around its circumference and a weight attached to the end of the cord. As the turbine rotates the weight is raised in the surroundings and the product of the force (due to the gravitational pull on the weight) and the distance through which the weight is raised gives the work transfer.

In both the non-flow and the steady-flow cases the work transfer is a product of force times distance (and force is a product of pressure times area.)

To take the non-flow case again, the easiest way to see this is to consider a piston of area A under the action of a pressure p in which there is a piston movement (and therefore system boundary movement) of $\mathrm{d}L$.

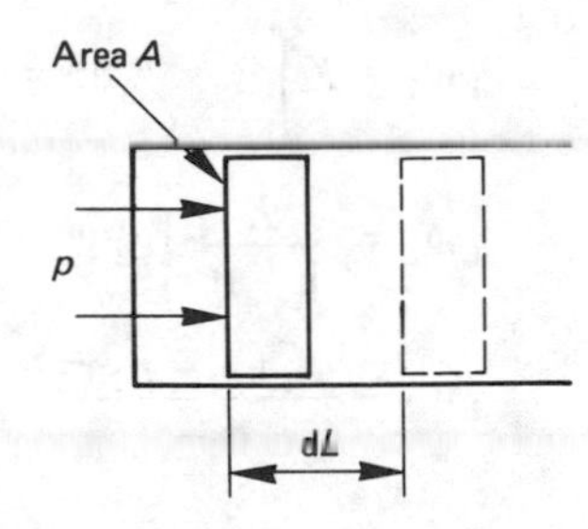

Elemental work $\delta W = F\,\mathrm{d}L = \dfrac{F}{A} A\,\mathrm{d}L = p\,\mathrm{d}V$, where $\mathrm{d}V$ is the volume displaced by the system boundary.

Note that the quantity $p\,\mathrm{d}V$ cannot be integrated unless the process is thermodynamically reversible. That is, no pressure gradients can exist at any points in the system (implying no turbulence). In real, irreversible flow the work can be calculated only from a knowledge (if one exists) of other quantities such as heat transfer and changes in internal energy using the first law or by experimental means.

However, reversible work transfer, like all ideal changes, is a very useful concept with which to compare reality.

Work is considered to be positive when delivered *by* the system *to* the surroundings and negative in the opposite sense.

Work is an interaction and therefore *not* a system property and elemental work is depicted with the use of δ and not d, the latter being exclusively reserved for property changes.

3.2 Worked Examples

Example 3.1

Show that the work done by the expansion of unit mass of a fluid from a state p_1, v_1 to another state p_2, v_2 according to the law

$$pv^n = \text{constant}$$

is given by

$$_1w_2 = \frac{p_1v_1 - p_2v_2}{n - 1}.$$

The work done in expanding gas from an initial volume of 0.1 m³ to 0.4 m³ is 80 kJ when the initial and final pressures are 10 bar and 2 bar respectively.

What is the value of n if the above law holds good for the expansion? How does w vary with n?

Answer

$$_1w_2 = \int_{v_1}^{v_2} \delta w = \int_{v_1}^{v_2} p\,dv$$

where δw is the elemental work,
p = variable pressure,
v_1, v_2 are initial and final specific volumes;

$$_1w_2 = \int_{v_1}^{v_2} \frac{K\,dv}{v^n} \qquad \text{(if } pv^n = K \text{ a constant);}$$

$$_1w_2 = K\int_{v_1}^{v_2} v^{-n}\,dv = \frac{K}{1-n}\left[v^{1-n}\right]_{v_1}^{v_2};$$

$$_1w_2 = \frac{K}{1-n}\left[v_2^{1-n} - v_1^{1-n}\right] = \frac{p_2v_2^n\,v_2^{1-n} - p_1v_1^n\,v_1^{1-n}}{1-n};$$

$$_1w_2 = \frac{p_1v_1 - p_2v_2}{n-1}.$$

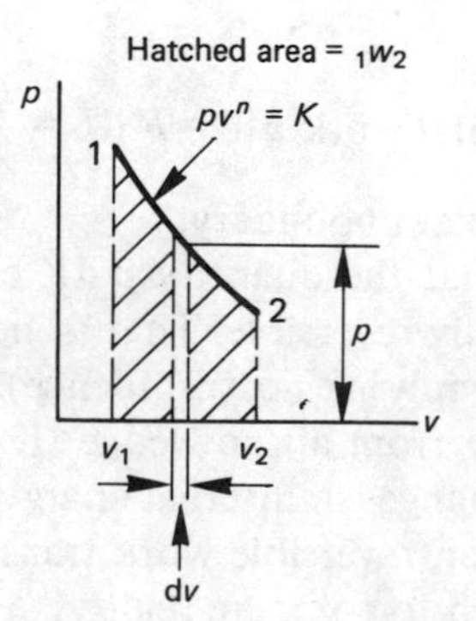

Notes

(a) The nomenclature shows $_1w_2$ as work occurring from state 1 to state 2 (i.e. 1 first and 2 last).

(b) $_2w_1 = -{}_1w_2$ by definition (the sense of w is vital).

(c) We can never have w_1 or w_2 i.e. work at a point (work is not a property but an interaction at a boundary).
(d) On a p–v field, expansion must be represented by a *full* line $1 \rightarrow 2$ representing a reversible process in which all intermediate states are known (unlike an irreversible process from a known state 1 to a known state 2 between which *there is no known path* and for which we can draw only an indeterminate dotted line).
(e) For the reversible case, ${}_1w_2$ is numerically given by the area under the curve as shown in the diagram.

In the gas expansion

$$n - 1 = \frac{p_1 v_1 - p_2 v_2}{{}_1w_2}$$

or

$$n = 1 + \left[\frac{p_1 v_1 - p_2 v_2}{{}_1w_2}\right]$$

or

$$n = 1 + \frac{\left[\left(1000\ \frac{\text{kN}}{\text{m}^2} \times 0.1\ \text{m}^3\right) - \left(200\ \frac{\text{kN}}{\text{m}^2} \times 0.4\ \text{m}^3\right)\right]}{80\ \text{kJ}} \quad \left[\frac{\text{kJ}}{\text{kN m}}\right]$$

$$= 1 + \left(\frac{100 - 80}{80}\right)$$

$$= 1.25.$$

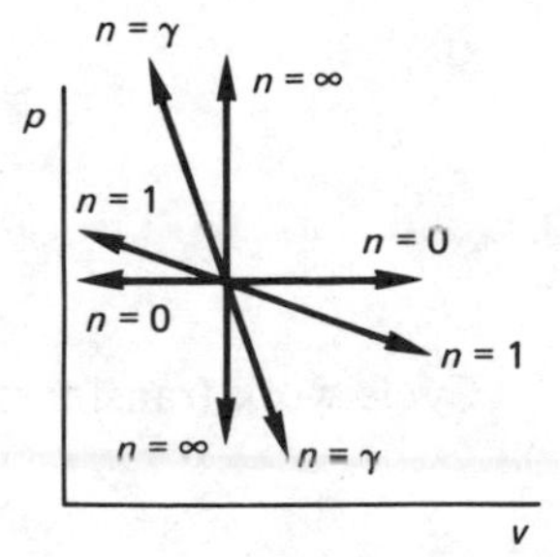

(f) As n decreases w increases. The graph depicts the following:

(i) $n = \infty$ for pressure increasing or decreasing and if

$$pv^{\infty} = \text{K}_1$$

then

$$p^{1/\infty} v = \text{K}_2,$$

i.e. v is constant (isochore or vertical line for which the area under is zero $\rightarrow$ no work)
(since $p^{1/\infty} = p^0 = 1$).

(ii) For $n = 0$, $pv^0 = \text{K}$ or p = constant (isobaric process).
(iii) If $n = \gamma$ we have a reversible adiabatic (see later).
(iv) For $n = 1$, $pv = \text{K}$ we have a hyperbolic process. (Note that in the special case of a perfect gas where $pv = RT$ this implies an isothermal process as well.)

Example 3.2

Unit mass of fluid undergoes the following cycle of operations in order:

(a) Isochoric compression from p_1, v_1 to p_2.
(b) Reversible adiabatic expansion (pv^{γ} = constant) to the original pressure.

(c) Isobaric energy rejection to p_1, v_1.

Given that $p_1 = 1$ bar, $p_2 = 5$ bar, $v_1 = 0.827$ m³/kg, calculate the cyclic work transfer if $\gamma = 1.4$.

Answer

Draw a picture of the cycle on the p–v field.

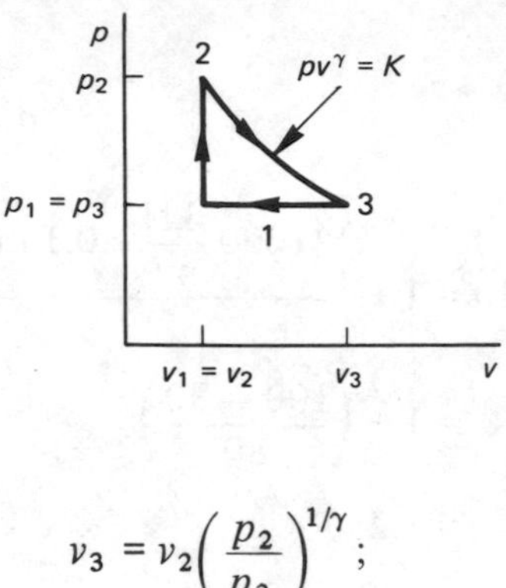

$$v_3 = v_2\left(\frac{p_2}{p_3}\right)^{1/\gamma};$$

$$v_3 = 0.827\ \frac{\text{m}^3}{\text{kg}}\,(5)^{0.714};$$

$$v_3 = 2.6096\ \frac{\text{m}^3}{\text{kg}}.$$

$$\text{Cycle work transfer} = \Sigma w = {}_1w_2 + {}_2w_3 + {}_3w_1$$

$$= 0 + \int_{v_2}^{v_3} p\,\mathrm{d}v + \int_{v_3}^{v_1} p\,\mathrm{d}v$$

$$= 0 + \frac{p_2v_2 - p_3v_3}{\gamma - 1} + p\int_{v_3}^{v_1} \mathrm{d}v$$

$$= 0 + \frac{p_2v_2 - p_2v_2}{\gamma - 1} + p_3\,(v_1 - v_3);$$

$$\therefore\ \Sigma w = \frac{\left(500\ \frac{\text{kN}}{\text{m}^2} \times 0.827\ \frac{\text{m}^3}{\text{kg}}\right) - \left(100\ \frac{\text{kN}}{\text{m}^2} \times 2.6096\ \frac{\text{m}^3}{\text{kg}}\right)}{0.4} + 100\ \frac{\text{kN}}{\text{m}^2}\,(0.827 - 2.6096)\ \frac{\text{m}^3}{\text{kg}}$$

$$= 381.35 - 178.3 = 203.05\ \frac{\text{kJ}}{\text{kg}}.$$

Example 3.3

Reconsider the expansion of example 3.1, this time with the addition of a suction process at constant pressure p_1 to the start of expansion at p_1, v_1 and an exhaust process at constant pressure from the end of expansion p_2, v_2 as given in the diagram at the top of the next page.

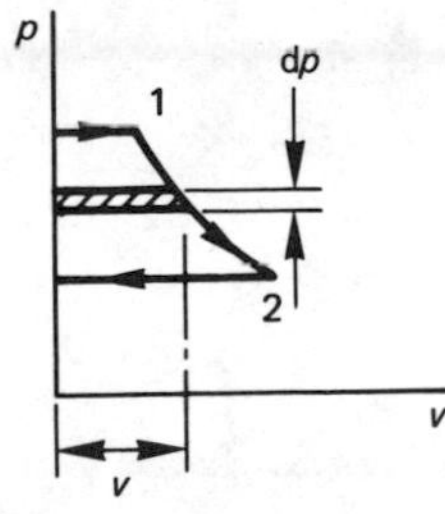

Calculate the work transfer for these three operations together and then show that the result is numerically given by the value of $\int_{p_2}^{p_1} v\, dp$.

Answer

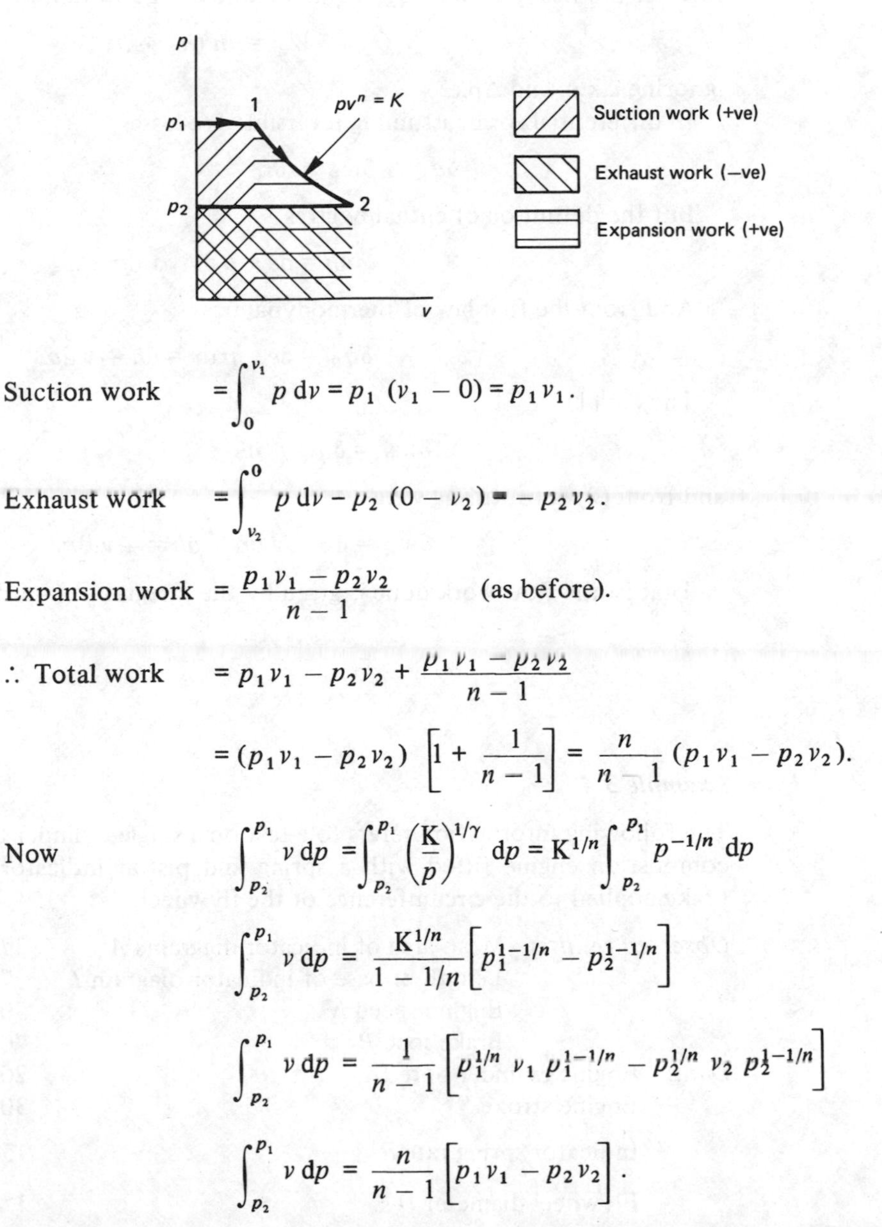

Suction work $= \int_0^{v_1} p\, dv = p_1 (v_1 - 0) = p_1 v_1 .$

Exhaust work $= \int_{v_2}^{0} p\, dv = p_2 (0 - v_2) = - p_2 v_2 .$

Expansion work $= \dfrac{p_1 v_1 - p_2 v_2}{n - 1}$ (as before).

$\therefore$ Total work $= p_1 v_1 - p_2 v_2 + \dfrac{p_1 v_1 - p_2 v_2}{n - 1}$

$$= (p_1 v_1 - p_2 v_2) \left[1 + \frac{1}{n - 1}\right] = \frac{n}{n - 1} (p_1 v_1 - p_2 v_2).$$

Now $\displaystyle\int_{p_2}^{p_1} v\, dp = \int_{p_2}^{p_1} \left(\frac{K}{p}\right)^{1/\gamma} dp = K^{1/n} \int_{p_2}^{p_1} p^{-1/n}\, dp$

$$\int_{p_2}^{p_1} v\, dp = \frac{K^{1/n}}{1 - 1/n} \left[p_1^{1-1/n} - p_2^{1-1/n}\right]$$

$$\int_{p_2}^{p_1} v\, dp = \frac{1}{n - 1} \left[p_1^{1/n}\, v_1\, p_1^{1-1/n} - p_2^{1/n}\, v_2\, p_2^{1-1/n}\right]$$

$$\int_{p_2}^{p_1} v\, dp = \frac{n}{n - 1} \left[p_1 v_1 - p_2 v_2\right] .$$

Note the significance of this result by referring to the next figure (overleaf).

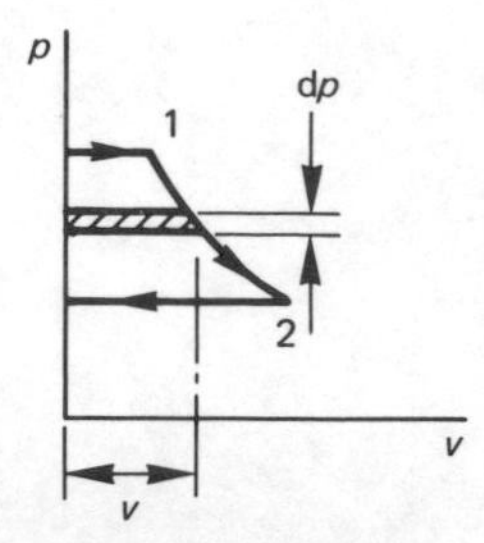

The combined process is a *flow* process (whereas the original process of expansion was a *non*-flow process).

A typical application of the flow case is flow through a fan (in the reverse sense to that shown here) or through a reciprocating expander (in the same sense).

Alternatively, we may derive the same result by reference to the differential form of the steady-flow energy equation and the definition of enthalpy as follows:

$$ {}_1\dot{Q}_2 - {}_1\dot{W}_2 = \dot{m}\,(h_2 - h_1), $$

ignoring Δk.e. and Δp.e.

In differential form, assuming reversible processes,

$$ \delta q_{\mathrm{R}} - \delta w_{\mathrm{R}} = \mathrm{d}h. \qquad \ldots (1) $$

But the definition of enthalpy gives

$$ \mathrm{d}h = \mathrm{d}e + p\,\mathrm{d}v + v\,\mathrm{d}p. $$

And from the first law of thermodynamics

$$ \delta q_{\mathrm{R}} = \mathrm{d}e + p\,\mathrm{d}v = \mathrm{d}h - v\,\mathrm{d}p. \qquad \ldots (2) $$

Thus in (1)

$$ \delta w_{\mathrm{R}} = \delta q_{\mathrm{R}} - \mathrm{d}h, $$

and from (2) substituting we get

$$ \delta w_{\mathrm{R}} = \mathrm{d}h - v\,\mathrm{d}p - \mathrm{d}h = -\,v\,\mathrm{d}p. $$

That is, the flow work done is given by the integral of $(-\,v\,\mathrm{d}p)$.

***Example** 3.4*

The following information refers to a test on a single-cylinder four-stroke internal-combustion engine fitted with a spring and piston indicator and a simple rope brake applied to the circumference of the flywheel.

Observed results:	Mean area of indicator diagrams A	370 mm^2
	Length of base of indicator diagram L	70 mm
	Engine speed N	500 rev/min
	Brake load P	700 N
Data:	Engine cylinder bore D_{E}	200 mm
	Engine stroke S	300 mm
	Indicator spring rate R	$150 \times 10^6\ \dfrac{\mathrm{N}}{\mathrm{m}^3}$
	Flywheel diameter D_{F}	1.22 m

Calculate the indicator power and the brake power of the engine.

Answer

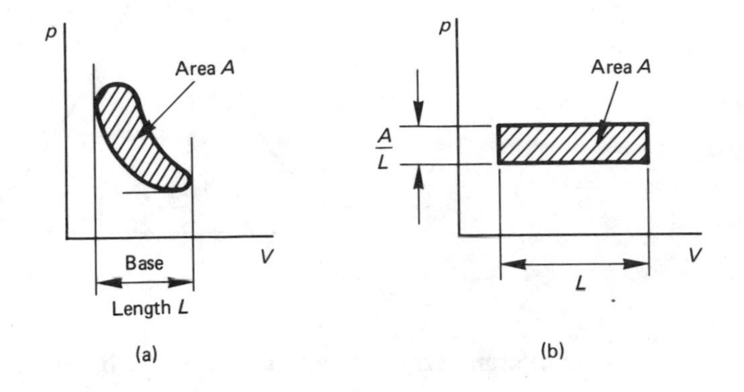

Mean effective pressure for diagram $= \dfrac{A}{L}$ (as in (b)) $= p_M$;

$$p_M = \frac{370\ \text{mm}^2}{70\ \text{mm}} \times 150 \times 10^6\ \frac{\text{N}}{\text{m}^3}\left[\frac{\text{m}}{10^3\ \text{mm}}\right]\left[\frac{\text{kN}}{10^3\ \text{N}}\right] = 792.9\ \frac{\text{kN}}{\text{m}^2}$$

(i.e. that constant pressure acting over the entire stroke which gives the same power as the actual diagram on the left).
Swept volume of engine in unit time = $\dot{V}_S$.

$$\dot{V}_S = \frac{\pi D_E^2}{4} \times S \times \frac{N}{2} \qquad \text{(one suction stroke every two revolutions of crankshaft)}$$

$$= \frac{\pi \times 200^2}{4}\ \text{mm}^2 \times 300\ \frac{\text{mm}}{\text{stroke}} \times 250\ \frac{\text{stroke}}{\text{min}}\left[\frac{\text{m}^3}{10^9\ \text{mm}^3}\right]$$

$$\dot{V}_S = 2.356\ \frac{\text{m}^3}{\text{min}}.$$

$$\text{Indicated power} = \dot{W}_I = p_M \dot{V}_S = 792.9\ \frac{\text{kN}}{\text{m}^2} \times 2.356\ \frac{\text{m}^3}{\text{mm}}\left[\frac{\text{min}}{60\ \text{s}}\right]\left[\frac{\text{kW s}}{\text{kN m}}\right] = 31.13\ \text{kW}.$$

Brake power = torque x angular velocity of flywheel;
$\dot{W}_B$ = brake load x flywheel radius x angular velocity

$$= 700\ \text{N} \times 0.61\ \text{m} \times 500\ \frac{\text{rev}}{\text{min}}\left[\frac{2\pi\ \text{rad}}{\text{rev}}\right]\left[\frac{\text{min}}{60\ \text{s}}\right]\left[\frac{\text{kW s}}{10^3\ \text{N m}}\right].$$

$\dot{W}_B$ = 22.36 kW.

Example 3.5

The indicated power output of a high-speed single cylinder compression-ignition engine is to be determined by means of a high-speed engine indicator system using an electrical transducer in the cylinder and an electronic crank-angle marker.

The resulting cylinder-pressure–crank-angle diagram is shown at the top of the next page with salient values of pressure–crank-angle listed in the table below it.

Calculate the indicated power if the engine is a four-stroke cycle unit and rotates at 2000 rev min^{-1}.

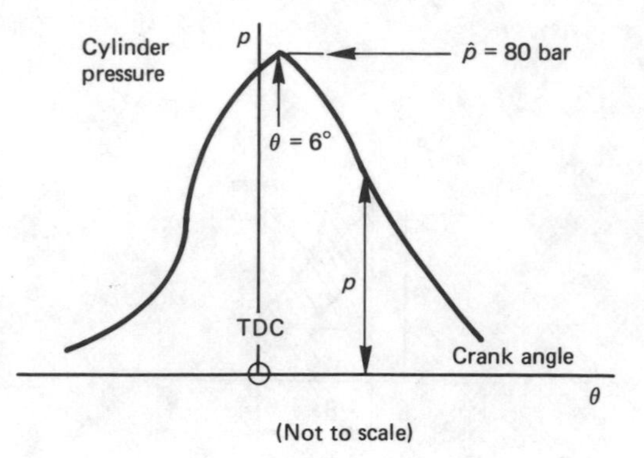

Data: crank arm 145 mm, con-rod length 580 mm, piston diameter 120 mm.

$\theta/°$	p/bar	$\theta/°$	p/bar
30	68.5	210	3.5
60	40.0	240	6.5
90	20.5	270	11.5
120	14.0	300	19.0
150	10.5	330	52.5
180	1.0	360	78.5 (t.d.c.)

Answer

Consider the engine-cylinder–crank-angle configuration shown below.

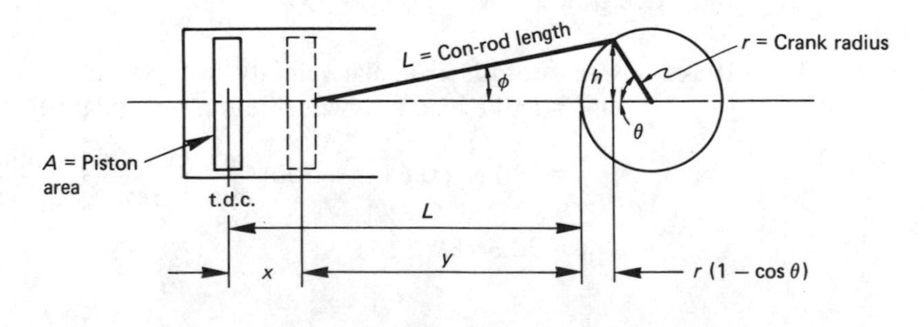

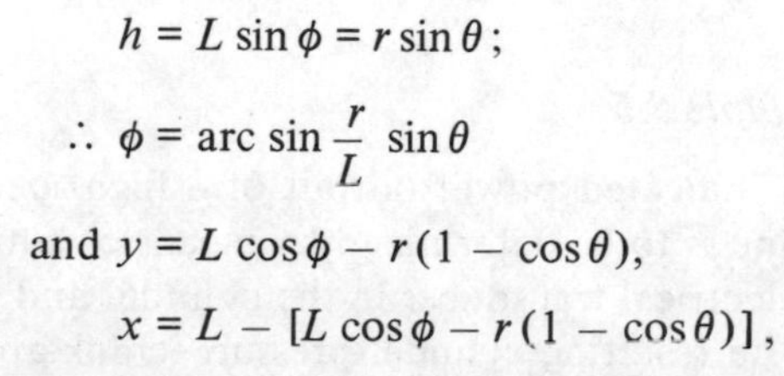

$$h = L \sin\phi = r \sin\theta ;$$

$$\therefore\ \phi = \text{arc sin} \frac{r}{L} \sin\theta$$

$$\text{and } y = L \cos\phi - r(1 - \cos\theta),$$

$$x = L - [L \cos\phi - r(1 - \cos\theta)],$$

where x is the displacement corresponding to the change in the engine swept volume for a crank angle movement of θ. (Note that the piston area A is constant throughout.)

Thus we may now show derived results as in the table.

θ	x	p/bar	θ	x	p/bar
30	0.239	68.5	210	2.751	3.5
60	0.863	40.0	240	2.313	6.5
90	1.634	20.5	270	1.634	11.5
120	2.313	14.0	300	0.863	19.0
150	2.751	10.5	330	0.239	52.5
180	2.900	1.0	360	0.0	78.5 (t.d.c.)

Note that $\hat{p}$ occurs at a crank angle of 6° and is 80 bar; also that $x = 2.9$ corresponds to the full engine swept volume V_S (i.e. twice the crank throw).

$$V_S = \frac{\pi \times 120^2}{4}\ \text{mm}^2 \times 290\ \text{mm} \left[\frac{\text{m}^3}{10^9\ \text{mm}^3}\right] = 3.28 \times 10^{-3}\ \text{m}^3.$$

If the corresponding p–V_S diagram is now drawn (a sketch is given here) and the area measured by some convenient method (e.g. planimeter) then

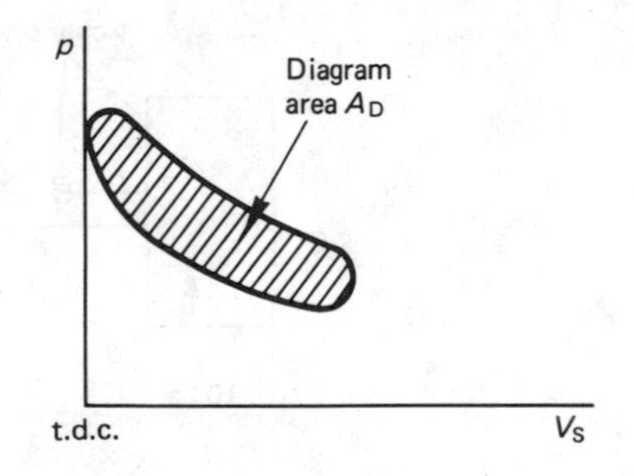

$A_D = 1000\ \text{mm}^2$ approximately on a diagram where
vertical axis is 1 mm ≡ 1 bar,
horizontal axis is 1 mm ≡ 0.025 V_S.

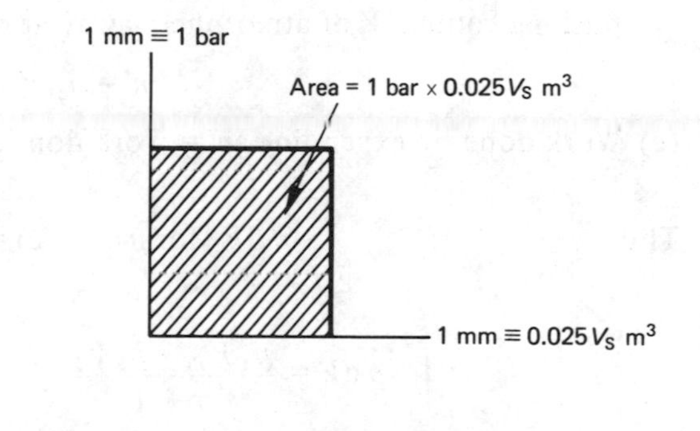

$$W_I = 1000\ \text{mm}^2 \times \frac{1\ \text{bar} \times 0.025 \times 3.28 \times 10^{-3}\ \text{m}^3}{\text{mm}^2\ \text{area}} \left[\frac{10^2\ \text{kN}}{\text{bar m}^2}\right]$$

$$= 2.828\ \text{kJ per cycle.}$$

Thus $$\dot{W}_I = 2.828\ \frac{\text{kJ}}{\text{cycle}} \times 1000\ \frac{\text{cycle}}{\text{min}} \left[\frac{\text{min}}{60\ \text{s}}\right] \left[\frac{\text{kW s}}{\text{kJ}}\right] = 47.13\ \text{kW}$$

(since these are 1000 cycles per minute for a four-stroke engine at 2000 rev min^{-1}).

The calculation of diagram area here has been carried out by drawing out on 1-millimetre graph paper and counting the squares, giving an answer of only rough accuracy.

Example 3.6

Air at 10 bar in the chamber of a compressed-air rifle expands reversibly and adiabatically to twice its original volume as the slug accelerates along the bore of the barrel. The mass of the slug is 1 g and it leaves the bore with a muzzle velocity of 100 m/s.

(a) Find the kinetic energy of the slug leaving the muzzle.
(b) Write a symbolic expression for the work done by the slug on the atmosphere.
(c) Assuming that the slug moves without friction and including the work in (b) above, determine the necessary reservoir capacity. Atmospheric pressure is 100 kN/m^2.

You may assume that the expansion of the air follows the law pv^γ = constant where $\gamma = 1.4$ for air.

Answer

Assuming the kinetic energy of the air is small, then:

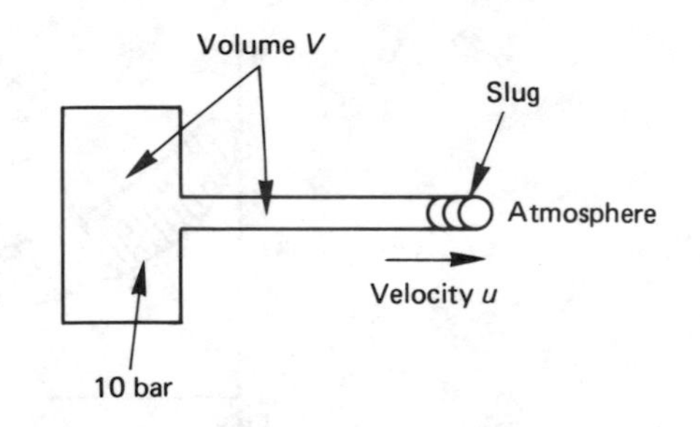

(a) $\text{k.e.} = \frac{1}{2}mu^2 = 0.5 \times 1\text{ g} \times 100^2\ \frac{\text{m}^2}{\text{s}^2}\left[\frac{\text{J}}{\text{N m}}\right]\left[\frac{\text{N s}^2}{\text{kg m}}\right]\left[\frac{\text{kg}}{10^3\text{ g}}\right] = 5\text{ J}.$

(b) Gas expansion down the barrel is non-flow.

In the absence of friction $\delta W = p\,\mathrm{d}V$ so that the work done by the slug pushing volume V of atmosphere away at constant pressure p_{at} is

$$W = p_{at}V.$$

(c) Work done by expanding air = work done on slug + work done on atmosphere.

Thus
$$\int_1^2 p\,\mathrm{d}V\ (\text{for } pV^\gamma = \text{c}) = p_{at}V + \text{k.e.}$$

$$\int_1^2 p\,\mathrm{d}V = \frac{p_1V_1 - p_2V_2}{\gamma - 1} \qquad \text{(see example 3.1).}$$

$$p_2 = p_1\left(\frac{V_1}{V_2}\right)^\gamma = 10\text{ bar}\left(\frac{1}{2}\right)^{1.4} = 3.79\text{ bar}.$$

Thus
$$\frac{[(10 \times 1) - (3.79 \times 2)]\ V\text{ bar}}{0.4} = 1\text{ bar} \times V + 5\text{ J},$$

$$\therefore \left[\left(\frac{10 - 7.58}{0.4}\right) - 1\right]\text{ bar} \times V = 5\text{ J};$$

or
$$V = \frac{5\text{ J}}{5.05\text{ bar}}\left[\frac{\text{N m}}{\text{J}}\right]\left[\frac{\text{bar m}^2}{10^5\text{ N}}\right] = 9.9 \times 10^{-6}\text{ m}^3$$
$$= 9.9\text{ ml}.$$

Example 3.7

A bicycle pump has a total stroke of 25 cm and is used to pump air into a tyre against a pressure of 3.5 bar.

Calculate the length of stroke necessary before air enters the tyre when the piston is pushed in

(a) slowly.
(b) rapidly.

Assume atmospheric pressure is 1 bar.

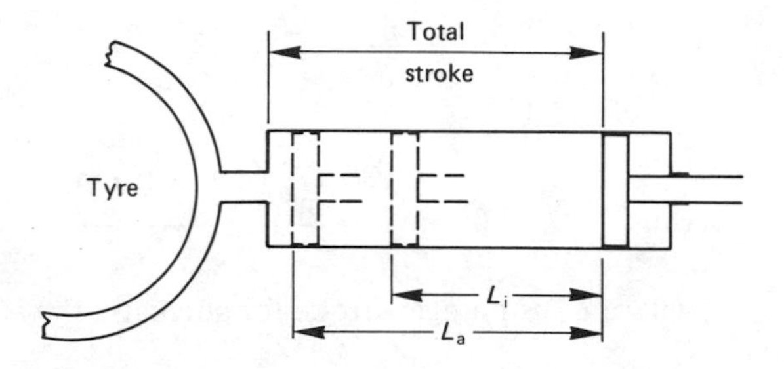

Answer

(a) Slow movement implies reversible, isothermal flow of air.
Let the plunger movement be L_i in this case.
Let the plunger cross-sectional area be A m^2.

$$W_i = \int_1^2 p\,dV = \int_{V_1}^{V_2} RT\frac{dV}{V} = RT\ln\frac{V_2}{V_1} = p_1V_1\ln\frac{V_2}{V_1}$$

since air may be treated as a perfect gas.

Thus $$W_i = 100\,\frac{\text{kN}}{\text{m}^2}\,(A \times 0.25)\text{ m}^3 \ln\frac{A\,(0.25 - L_i)\text{ m}^3}{A\,(0.25)\text{ m}^3}$$

$$= 25\,A\ln\left(\frac{0.25 - L_i}{0.25}\right)\text{ kJ.}$$

However, $$W_i = \int_{V_1}^{V_2} p\,dV = \int_{p_1}^{p_2} p\left(-\frac{K}{p^2}\right)dp$$

since $pV = \text{K}$ and $V = \dfrac{K}{p}$.

Thus $dV = -\dfrac{K}{p^2}\,dp$,

$$W_i = -\text{K}\int_{p_1}^{p_2}\frac{dp}{p} = -K\Big[\ln p\Big]_{p_1}^{p_2} = p_1V_1\ln\frac{p_1}{p_2},$$

or $$W_i = 100\,\frac{\text{kN}}{\text{m}^2}\,(A \times 0.25)\text{ m}^3\ln\frac{1}{3.5} = -\,31.319A\text{ kJ}$$

Thus, equating the two experiences for W_i and cancelling A out,

$$25\ln\left(\frac{0.25 - L_i}{0.25}\right) = -\,31.319$$

$$\text{or} \qquad \frac{0.25}{0.25 - L_i} = \text{antiln}\ \frac{31.319}{25} = 3.5,$$

$$\text{or} \qquad L_i = 0.179 \text{ m} \quad (17.9 \text{ cm}).$$

For adiabatic flow there is no heat transfer.

$$\textit{First Law:}\ -W_a = E_2 - E_1 = m\,c_v\,(T_2 - T_1) = \frac{mR}{\gamma - 1}(T_2 - T_1)$$

$$= \frac{p_2 V_2 - p_1 V_1}{\gamma - 1};$$

$$\text{or} \qquad W_a = \frac{p_1 V_1 - p_2 V_2}{\gamma - 1};$$

$$\text{thus} \qquad W_a = \frac{100\ \dfrac{\text{kN}}{\text{m}^2}(A \times 0.25)\ \text{m}^3 - 350\ \dfrac{\text{kN}}{\text{m}^2}\ A\,(0.25 - L_a)}{0.4},$$

where L_a is plunger stroke for adiabatic flow.

$$\text{Now} \qquad W_a = \int_{V_1}^{V_2} p\,\mathrm{d}V$$

$$\text{and for adiabatic flow} \qquad pV^\gamma = K$$

$$\text{or} \qquad V = \left(\frac{K}{p}\right)^{1/\gamma}$$

$$\text{so that} \qquad \mathrm{d}V = K^{1/\gamma}\left(-\frac{1}{\gamma}\right)p^{-1/\gamma - 1}\ \mathrm{d}p\ .$$

Thus

$$W_a = \int_1^{3.5} p\left(-\frac{1}{\gamma}\right)\mathrm{K}^{1/\gamma}p^{-1/\gamma - 1}\ \mathrm{d}p = -\frac{\mathrm{K}^{1/\gamma}}{\gamma}\int_1^{3.5} p^{-1/\gamma}\ \mathrm{d}p.$$

Thus

$$W_a = -\frac{\mathrm{K}^{1/\gamma}}{\gamma}\left[p^{1-1/\gamma}\right]_1^{3.5}\frac{1}{1 - 1/\gamma} = -\frac{\mathrm{K}^{1/\gamma}}{\gamma - 1}\left[3.5^{0.286} - 1\right]$$

$$= -\frac{p_1^{1/\gamma}V_1}{\gamma - 1}(0.431) = -\frac{100\ \dfrac{\text{kN}}{\text{m}^2}(A \times 0.25)\ \text{m}^2}{0.4}(0.431) = -7.227A \text{ kJ}$$

and equating the two values for W_a,

$$-7.227\,A = 62.5A - 218.75A + 875AL_a, \text{ and cancelling } A,$$

$$L_a = \frac{149.023}{875}\ \text{m} = 0.170 \text{ m } (17.0 \text{ cm}).$$

Exercises

(all for non-flow processes)

1 Calculate the work done by a gas which expands hyperbolically (pV = const.) from 8 bar to 1 bar when the initial volume is 1 m^3. (1663.6 kJ)

2 In a gas cycle there are three operations in order:

(a) Reversible adiabatic compression from 1 bar, 0.3 m^3 to 6 bar ($\gamma = 1.4$).
(b) Reversible hyperbolic expansion (pV = constant) back to 1 bar.
(c) Isobaric cooling back to the initial state.

Calculate the cyclic work transfer. (19.9 kJ)

3 A gas obeys the equation $p = \dfrac{RT}{v - b} - \dfrac{a}{v^2}$ where R, a and b are constants and T is 288 K. If T is constant and the gas expands from 10 bar, 0.8 m^3 to 2 bar, 1.6 m^3, calculate the work done per kg gas. $R = 0.287$ kJ/(kg K). (359.5 kJ/kg)

4 A mass of air occupying 0.5 m^3 at 2 bar and 200°C is compressed reversibly and adiabatically to 5 bar and then 60 kJ of heat are added isobarically. Calculate the net work transfer for these two processes.

Calculate also the index of reversible polytropic expansion n which will return the fluid to its initial state. Calculate the work done in this process and the cyclic work transfer overall. (−14.94 kJ, 1.728, 64.7 kJ, 50.23 kJ)

4 Properties of the Working Fluid

4.1 Vapours

There are six fundamental fluid properties of importance, as follows:

(a) pressure p;
(b) temperature T;
(c) volume V;
(d) internal energy E;
(e) enthalpy H;
(f) entropy S.

Entropy is dealt with separately in Chapter 9 when the second law of thermodynamics is considered.

A clear understanding of the manner in which the value of any property is established is possible only when a 'picture' of events is given. This is the plot of any one property against another, e.g. p/V, T/E, p/H, etc.

Two such pictures are given here, namely T/v, where $v = V/m$, i.e. specific volume, and T/h, where $h = H/m$, i.e. the specific enthalpy.

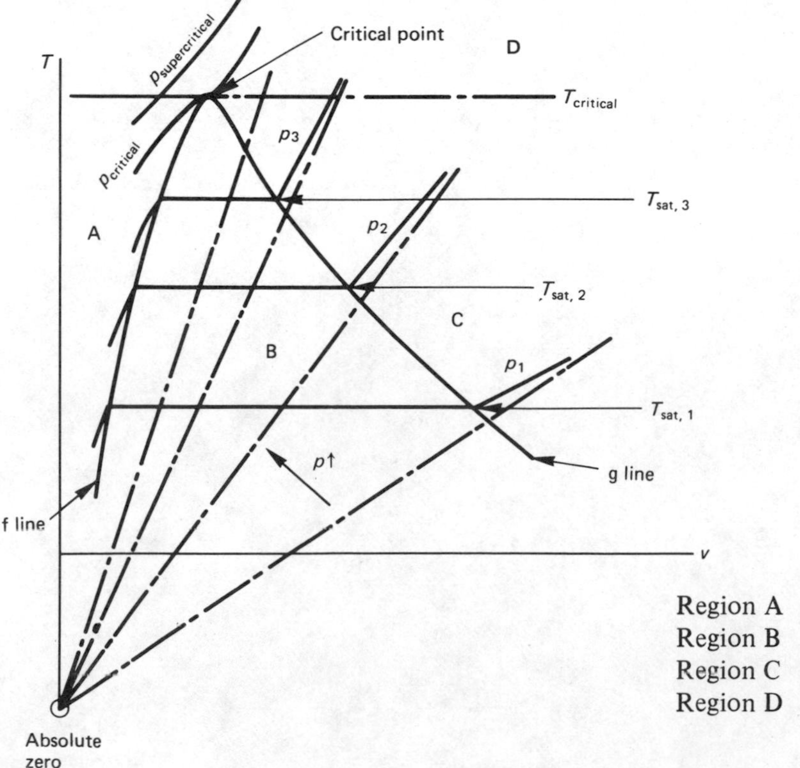

Region A Subcooled liquid
Region B 'Wet' vapour
Region C Superheated vapour
Region D Gas

The liquid saturation line (f line) and the vapour saturation line (g line) merge at the critical point. The departure of the isobars from the f line has been exaggerated for the sake of clarity. Isobars are asymptotic to straight lines radiating from the point of zero absolute temperature. Pressure and temperature are *not* independent in the 'wet' vapour region.

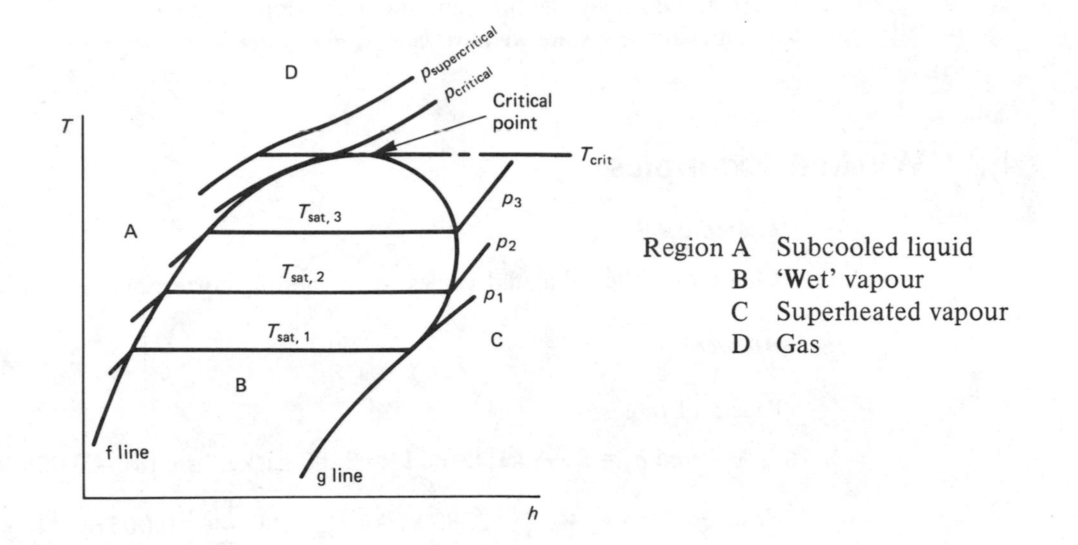

Note that $\mathrm{d}T/\mathrm{d}h$ is zero in the wet-vapour region, or taking the reciprocal $\mathrm{d}h/\mathrm{d}T$ is infinite and thus by definition c_p is also infinite.

Thus property evaluation may be divided into *two* distinct parts, as follows:

(a) Properties below the critical temperature for which tabulated values are needed.
(b) Properties above the critical temperature for which an equation, namely $pV = mRT$, exists and direct calculation is possible.

Properties Below the Critical Temperature

Reference to a typical page from Rogers and Mayhew tables for wet steam contains the following columns:

$\frac{p}{\text{bar}}$	$\frac{T}{^\circ\text{C}}$	$\frac{v_g}{\text{m}^3/\text{kg}}$	$\frac{u_f}{\text{kJ/kg}}$	$\frac{u_g}{\text{kJ/kg}}$	$\frac{h_f}{\text{kJ/kg}}$	$\frac{h_{fg}}{\text{kJ/kg}}$	$\frac{h_g}{\text{kJ/kg}}$

(Note that all reference to entropy S is omitted at this stage.)

Because of the nomenclature in use in this book the following revised symbols are used:

$\frac{p}{\text{bar}}$	$\frac{T}{^\circ\text{C}}$	$\frac{v_g}{\text{m}^3/\text{kg}}$	$\frac{e_f}{\text{kJ/kg}}$	$\frac{e_g}{\text{kJ/kg}}$	$\frac{h_f}{\text{kJ/kg}}$	$\frac{h_{fg}}{\text{kJ/kg}}$	$\frac{h_g}{\text{kJ/kg}}$

Saturated vapour properties have the suffix g and saturated liquid properties the suffix f as in the state diagrams above.

Values of v_f are on page 10 of the tables, with temperature as the independent variable since the latter is the important quantity, pressure having negligible effect upon the value of specific volume in the compressed or subcooled liquid region.

Further reference shows that on page 2 temperature is the extreme left-hand column, i.e. the independent variable, pressure is the independent variable on pages 3 to 5 for wet vapour and on pages 6 to 9 for superheated vapour.

Values of e_{fg} are not given (to save on space) and subtraction is necessary (i.e. $e_{fg} = e_g - e_f$).

Note that saturated vapour values of the salient properties are requoted on pages 6 to 9 to allow direct interpolation between these states and the first of the quoted superheated states on these pages.

The first thing is to check that values of specific internal energy and specific enthalpy do obey the fundamental definition $h = e + pv$.

Thus at the g state we have $h_g = e_g + pv_g$.

4.2 Worked Examples

Example 4.1

Check that the tabulated values for e_g and h_g correspond at a pressure of 20 bar.

Answer

Page 5 of tables:

listed h_g = 2398 kJ; listed e_g = 2283 kJ/kg; listed v_g = 0.005 69 m³/kg.

Now
$$e_g + pv_g = 2283 \text{ kJ/kg} + \left(20\,200 \ \frac{\text{kN}}{\text{m}^2}\right)\left(0.005\,69 \ \frac{\text{m}^3}{\text{kg}}\right)\left[\frac{\text{kJ}}{\text{kN m}}\right]$$
$$= 2397.9 \text{ kJ/kg} \qquad \text{(cf. 2398 kJ/kg above).}$$

Note here that 202 bar has been replaced by 20 200 $\frac{\text{kN}}{\text{m}^2}$ by multiplying by 100 since there are 100 $\frac{\text{kN}}{\text{m}^2}$ in 1 bar and 1 bar is not a preferred SI unit (= 10^5 N/m²).

Correspondingly, any other value of h and e should agree in this way. Consider now wet steam of dryness x.

$$v = xv_g + (1 - x)\, v_f \triangleq xv_g$$

(accurate enough for most purposes since v_f is small).

Example 4.2

Find the values of e and h for wet steam at 0.8 bar, 0.75 dry (p. 3 of tables).

Answer

$$v_g = 2.087 \ \frac{\text{m}^3}{\text{kg}}; \ v = xv_g = 0.75 \times 2.087 \ \frac{\text{m}^3}{\text{kg}} = 1.565\,25 \ \frac{\text{m}^3}{\text{kg}}.$$

$h = h_f + xh_{fg}$ where $h_{fg} = h_g - h_f$ by definition;

$h = 392 + 0.75\,(2273) = 2096.75$ kJ/kg.

$$e = e_f + xe_{fg} = e_f + x\,(e_g - e_f)$$
$$= 392 + 0.75\,(2498 - 392) = 1971.5 \text{ kJ/kg.}$$

Check: $h - e = 125.25$ kJ/kg

and
$$pv = 80 \ \frac{\text{kN}}{\text{m}^2} \times 1.565\,25 \ \frac{\text{m}^3}{\text{kg}} = 125.22 \text{ kJ/kg.}$$

Example 4.3

All horizontal interpolations between columns may be assumed linear. However, vertical interpolation will *not* suffice in every case.

On page 6, at low pressure, specific volume is not linear vertically;

e.g. v at 0.3 bar, 100°C *is not given by* $\dfrac{v_{0.1,100} + v_{0.5,100}}{2}$;

i.e. by $\dfrac{17.2 + 3.42}{2} = 10.31\ m^3/kg$.

The graph of pressure against specific volume is not remotely linear in this region. Reference to the expression at the foot of page 11 will show the true linearity which is in the *product of p and v*. This is so because enthalpy is always linear vertically as well as horizontally and the expression shows a linear relationship between pv and h.

Thus, using the expression first for the above example,

$$h_{0.3,100} = \frac{h_{0.1,100} + h_{0.5,100}}{2} = \frac{2688 + 2683}{2} = 2685.5\ kJ/kg$$

and $$v_{0.3,\,100} = \frac{(0.3/1.3)\,(2685.5 - 1943)}{100 \times 0.3} = 5.7115\ m^3/kg.$$

Alternatively (p. 6 of tables),

$$(pv)_{0.1,\,100} = \left(10\ \frac{kN}{m^2} \times 17.2\ \frac{m^3}{kg}\right) = 172\ \frac{kJ}{kg},$$

$$(pv)_{0.5,\,100} = \left(50\ \frac{kN}{m^2} \times 3.42\ \frac{m^3}{kg}\right) = 171\ \frac{kJ}{kg}.$$

Thus $$(pv)_{0.3,\,100} = \frac{172 + 171}{2} = 171.5\ \frac{kJ}{kg}$$

and $$v_{0.3,\,100} = \frac{171.5\ kJ/kg}{30\ kN/m^2} = 5.717\ m^3/kg \qquad \text{(cf. 10.31 above!).}$$

Thus great care must be taken with v on page 6. Later on in the superheat tables the error is much smaller.

Example 4.4

Calculate v_f at 100 bar and 275 °C.

The important independent variable here, as explained earlier, is temperature since pressure has little effect on the value of the specific volume of compressed or subcooled liquid. (Note that values of v_f in the tables are quoted 100 times bigger – see head of column.)

On page 10,

$$v_f \text{ at } 270°C = 0.001\,302\ m^3/kg,$$

$$v_f \text{ at } 280°C = 0.001\,332\ m^3/kg.$$

Thus $$v_f \text{ at } 275\,°C = \frac{0.001\,302 + 0.001\,332}{2} = 0.001\,317\ m^3/kg.$$

Example 4.5

Complete the following table with explanation. Assume that for superheated steam

$$v_{spht} = v_g \times \frac{T_{spht}\ (\text{in K})}{T_{satn}\ (\text{in K})} \text{ approximately.}$$

	Fluid	p / bar	T / °C	x	v / m³/kg	e / kJ/kg	h / kJ/kg
(a)	H_2O	80		0.3			
(b)	H_2O	190			0.013 62		
(c)	H_2O		30		32.93		
(d)	Freon-12	0.168	−45				
(e)	NH_3	6.585	30				

(a) $0 < x < 1$ i.e. wet vapour between f and g lines with $T = T_{satn}$ constant. Page 5 at $p = 80$ bar; $T_{satn} = 295$ °C

$$v \triangleq x v_g = 0.3 \times 0.023\,52 = 0.007\,056 \text{ m}^3/\text{kg}.$$

$$e = e_f + x(e_g - e_f) = 1306 + 0.3(2570 - 1306),$$

$$e = 1685.2 \text{ kJ/kg}.$$

$$h = h_f + x(h_{fg}) = 1317 + 0.3(1441),$$

$$h = 1749.3 \text{ kJ/kg}.$$

(b) $v > v_g$ at 190 bar (page 5 $v_g = 0.006\,68$ m³/kg). Thus superheated steam. Thus use page 8 at $p = 190$ bar and note that $v = 0.013\,62 \frac{\text{m}^3}{\text{kg}}$ is at $T = 450$°C. Since steam is superheated x is non-applicable (NA).

$h_{190,450} = 3082$ kJ/kg.

$$e_{190,450} = h - pv = 3082 \frac{\text{kJ}}{\text{kg}} - \left(19\,000 \frac{\text{kN}}{\text{m}^2}\right)\left(0.013\,62 \frac{\text{m}^3}{\text{kg}}\right) = 2823.22 \text{ kJ/kg}.$$

(c) $v = 32.93$ m³/kg and $T = 30$ °C

gives a direct reading since

$v_g = 32.93$ m³/kg at $T = 30$ °C

(point lies on g line → satd vapour).

Thus $x = 1.0$;

$p = 0.042\,42$ bar;

$h = h_g = 2555.7$ kJ/kg;

$$e = e_g = h_g - pv_g = 2555.7 \frac{\text{kJ}}{\text{kg}} - \left(4.242 \frac{\text{kN}}{\text{m}^2}\right)\left(32.93 \frac{\text{m}^3}{\text{kg}}\right),$$

$$e = 2416.01 \text{ kJ/kg}.$$

(d) Freon-12 properties are listed on page 13 (no e or h_{fg} values directly).

Here $p = 0.168$ bar and $T = -45$ °C.
But $T_{sat} = -65$ °C.
Thus freon is superheated by 20 K (i.e. *between* 15 K and 30 K spht).

Thus x is NA (not applicable.)

$$v = v_g \frac{T_{spht}\ (K)}{T_{sat}\ (K)} \text{ approx.} = 0.8412\ \frac{m^3}{kg}\left(\frac{273 - 45}{273 - 65}\right) = 0.922\ \frac{m^3}{kg}.$$

h is given as being $\left(\frac{20 - 15}{30 - 15}\right)$ of the way between values listed at 15 K and 30 K superheat. (Degree of superheat is 20 K.)

Thus $\quad h = 165.70 + \frac{5}{15}\ (173.68 - 165.7) = 168.36\ kJ/kg$

and $\quad e = h - pv = 168.36\ \frac{kJ}{kg} - 16.8\ \frac{kN}{m^2}\left(0.922\ \frac{m^3}{kg}\right) = 152.87\ \frac{kJ}{kg}$.*

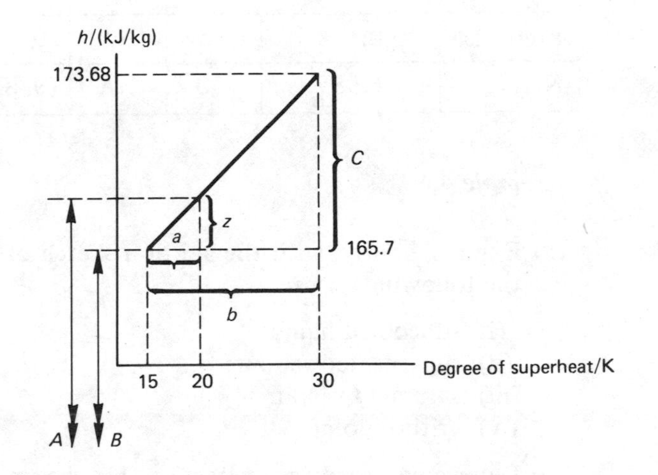

$A = B + z = B + \frac{a}{b}\ c$ (similar triangles).

Here $B = 165.7$ kJ/kg,
$a = 20 - 15 = 5$ K,
$b = 30 - 15 = 15$ K,
$c = (173.68 - 165.7)\ \frac{kJ}{kg}$,

as above.

(e) Ammonia properties are on page 12.

$$T_{sat}\ (\text{for } p = 6.585 \text{ bar}) = 12\ °C.$$

But $T = 30$ °C given. Thus 18 K superheat.
This lies between the g state and that at 50 K superheat.
Thus x is NA.

$$v = 0.1926\ \frac{m^3}{kg}\left(\frac{273 + 30}{273 + 12}\right) = 0.2048\ \frac{m^3}{kg} \text{ as in (d).}$$

$$h = h_g + \frac{18}{50}\ (h_{50\,K} - h_g) = 1456.1\ \frac{kJ}{kg} + \frac{18}{50}\ (1586 - 1456.1)\ \frac{kJ}{kg}$$

$$h = 1502.86\ \frac{kJ}{kg}.$$

*Note that this kind of interpolation can be represented as in the diagram.

$$e = h - pv = 1502.8\ \frac{\text{kJ}}{\text{kg}} - \left(658.5\ \frac{\text{kN}}{\text{m}^2}\right)\left(0.2048\ \frac{\text{m}^3}{\text{kg}}\right)$$

$$e = 1368.0\ \frac{\text{kJ}}{\text{kg}}.$$

The complete table with calculations in parentheses reads as shown here.

Fluid	$\frac{p}{\text{bar}}$	$\frac{T}{°\text{C}}$	x	$\frac{v}{\text{m}^3/\text{kg}}$	$\frac{e}{\text{kJ/kg}}$	$\frac{h}{\text{kJ/kg}}$
H_2O	80	(295)	0.3	(0.007 056)	(1685.2)	(1749.3)
H_2O	190	(450)	(NA)	0.013 62	(2823.22)	(3082.0)
H_2O	(0.042 42)	30	(1.0)	32.93	(2416.01)	(2555.7)
Freon-12	0.168	−45	(NA)	(0.922)	(152.87)	(168.36)
NH_3	6.585	30	(NA)	(0.2048)	(1368.0)	(1502.86)

Example 4.6

(a) Explain briefly, with the aid of a sketch of a property field, the meaning of the following terms:

(i) subcooled liquid;
(ii) superheated vapour;
(iii) saturated vapour;
(iv) critical point.

(b) Determine the dryness fraction, if a wet mixture, or temperature if a superheated vapour, of the following substances in the states given:

(i) water: pressure 10 bar, specific enthalpy 3588 kJ/kg;
(ii) water: pressure 10 bar, specific internal energy 1588 kJ/kg;
(iii) water: pressure 15 bar, specific volume 0.11 m^3/kg;
(iv) freon-12: temperature 15 °C, specific enthalpy 155 kJ/kg;
(v) mercury: pressure 5 bar, specific enthalpy 355 kJ/kg.

Answer

Refer in part (a) to the state diagrams at the start of this chapter.

(b) (i) $h = 3588\ \frac{\text{kJ}}{\text{kg}}$; and on page 4 at 10 bar $h_g = 2778\ \frac{\text{kJ}}{\text{kg}}$.

Thus superheated vapour:

on page 7 at $p = 10$ bar, $T = 500\ °\text{C} + \left(\frac{3588 - 3478}{3698 - 3478}\right)(600 - 500)\ \text{K}$

$$T = 550\ °\text{C}.$$

(ii) On page 4 at $p = 10$ bar, $e_g = 2584\ \frac{\text{kJ}}{\text{kg}}$, $e_f = 762\ \frac{\text{kJ}}{\text{kg}}$.

But $e = 1588\ \frac{\text{kJ}}{\text{kg}}$ given: thus wet vapour.

And $x = \frac{e - e_f}{e_g - e_f} = \frac{1588 - 762}{2584 - 762} = 0.453.$

(iii) On page 4 at p = 15 bar $v_g = 0.1317 \dfrac{m^3}{kg}$; but $v = 0.11 \dfrac{m^3}{kg}$.

Thus wet vapour and $x \triangleq \dfrac{v}{v_g} = \dfrac{0.11}{0.1317} = 0.8352$.

(iv) On page 13: $h = 155 \dfrac{kJ}{kg}$ given; but $h_g = 193.78 \dfrac{kJ}{kg}$; $e_f = 50.1 \dfrac{kJ}{kg}$.

Thus wet vapour and $x = \dfrac{h - h_f}{h_g - h_f} = \dfrac{155 - 50.1}{193.78 - 50.1} = 0.7301$.

(v) On page 4: p = 5 bar; $h_g = 352.78 \dfrac{kJ}{kg}$; but $h = 355 \dfrac{kJ}{kg}$ given.

Thus superheated vapour and in this case using information at the foot of page 14 and treating this vapour as a 'gas',

$h_{spht} = h_g + c_{p,spht}(T_{spht} - T_{sat})$ where $c_p = 0.1036 \dfrac{kJ}{kg\ K}$ given.

$$T_{spht} - T_{sat} = \text{degree of superheat} = \frac{h_{spht} - h_g}{c_{p,spht}}.$$

$$T_{spht} - T_{sat} = \frac{(355 - 352.78)\ \dfrac{kJ}{kg}}{0.1036\ \dfrac{kJ}{kg\ K}} = 21.43\ \text{K superheat.}$$

The use of this expression will be clarified further in the next section.

Exercises

1 Complete the following table:

Fluid	$\dfrac{p}{\text{bar}}$	$\dfrac{T}{°C}$	Regime or quality	$\dfrac{v}{m^3/kg}$	$\dfrac{e}{kJ/kg}$	$\dfrac{h}{kJ/kg}$
H_2O		230	0.6 dry			
H_2O	50				3000	
H_2O	100	130				
NH_3		−10	Saturated vapour			
Freon-12	2.61	5				

Answer:

Fluid	$\frac{p}{\text{bar}}$	$\frac{T}{°\text{C}}$	Regime or quality	$\frac{v}{\text{m}^3/\text{kg}}$	$\frac{e}{\text{kJ/kg}}$	$\frac{h}{\text{kJ/kg}}$
H_2O	28			0.0428 52	1957	2078.2
H_2O		450	Superheated	0.063 2		3316
H_2O			Subcooled liquid	0.001 07	546 (p. 4)*	546 (p. 4)*
NH_3	2.908			0.418 5	1311.3	1433.0
Freon-12			Superheated	†	†	191.73

*Treat fluid as though it is saturated liquid at 130 °C to find e and h.
†Cannot be estimated with information given on tables.

2 (a) Explain the meaning of the terms: saturated liquid, superheated vapour, dryness fraction.

(b) Make a careful sketch of the $T - v$ field for a pure substance such as H_2O or CO_2. Include the saturation boundary and a few typical isobars and indicate the compressed liquid, wet equilibrium mixture and superheated vapour regions.

(c) Determine whether H_2O in each of the following states is a compressed liquid, a superheated vapour or an equilibrium mixture of liquid and vapour.

(i) $p = 50$ bar, $T = 375$ °C;
(ii) $p = 10$ bar, $T = 55$ °C;
(iii) $T = 195$ °C, $v = 0.105$ m^3/kg.

In each case calculate e and h.

Answer

(c) (i) Superheated vapour: $e = 2858.5$ kJ/kg, $h = 3133$ kJ/kg;
(ii) compressed liquid: $e = 229.19$ kJ/kg, $h = 230.2$ kJ/kg;
(iii) wet vapour: $x = 0.7457$, $e = 2144.23$ kJ/kg, $h = 2291.65$ kJ/kg.

4.3 Gases

Properties Above Critical Temperature (Gases)

The laws of a perfect gas are summarised as follows:

$$pV = mRT = nR_0T$$

Where n = number of kmol of the substance of mass m and volume V at pressure p and temperature T,

T = absolute thermodynamic temperature in K (°C + 273),

and R_0 = universal gas constant = $8.3143 \dfrac{\text{kJ}}{\text{kmol K}}$ (p. 24 tables).

It then follows that $\dfrac{m}{n} = m_v$ = relative molar mass $\left(\dfrac{\text{kg}}{\text{kmol}}\right)$,

$$R = \frac{R_0}{m_v} = \text{specific gas constant} \left(\frac{\text{kJ}}{\text{kg K}}\right).$$

Furthermore, the two principal specific heat capacities are

$$\left.\begin{aligned} c_p &= \frac{dh}{dT} \\ c_v &= \frac{de}{dT} \end{aligned}\right\} \quad \text{assumed constant for a perfect gas}$$

and
$$c_p - c_v = \frac{R_0}{m_v} = R = \frac{\overline{c}_p - \overline{c}_v}{m_v},$$

where $\overline{c}_p$ and $\overline{c}_v$ are molar specific heat capacities

i.e.
$$\overline{c}_p = m_v c_p \text{ and } \overline{c}_v = m_v c_v.$$

It therefore follows that

$$dh = c_p \, dT$$

and integrating
$$(h_2 - h_1) = c_p \, (T_2 - T_1)$$

Correspondingly,
$$(e_2 - e_1) = c_v \, (T_2 - T_1).$$

Consequent upon the above relationships, the expressions for work transfer in a polytropic expansion or compression may be rewritten as follows:

When pv^n = constant

$$_1W_2 = \frac{p_1 v_1 - p_2 v_2}{n-1} = \frac{R(T_1 - T_2)}{n-1} \quad \text{for a non-flow process}$$

and
$$_1W_2 = \frac{nR(T_1 - T_2)}{n-1} \quad \text{for the flow process.}$$

Further consequences of the laws of a perfect gas are encountered in the next chapter when the first law of thermodynamics is introduced and also later on when the second law is introduced together with entropy.

4.4 Worked Examples

Example 4.7

Superheated mercury vapour may be assumed to behave like a perfect gas. Calculate the specific enthalpy, specific internal energy at 30 bar, 700 °C.

Answer

Page 14 of tables: h_g at 30 bar = 371.36 $\frac{\text{kJ}}{\text{kg}}$; $c_p = 0.1036 \frac{\text{kJ}}{\text{kg K}}$.

Thus $h = h_g + c_p \, (T - T_{\text{sat}})$

$$= 371.36 \frac{\text{kJ}}{\text{kg}} + 0.1036 \frac{\text{kJ}}{\text{kg K}} (700 - 630) \text{ K} = 378.612 \frac{\text{kJ}}{\text{kg K}};$$

$$v = v_g \frac{T_{\text{spht}}}{T_{\text{sat}}} = 0.012\,52 \frac{\text{m}^3}{\text{kg}} \times \frac{973}{903} = 0.013\,49 \frac{\text{m}^3}{\text{kg}};$$

$$e = h - pv = 378.16 \frac{\text{kJ}}{\text{kg}} - 3000 \frac{\text{kN}}{\text{m}^2} \left(0.013\,49 \frac{\text{m}^3}{\text{kg}}\right)$$

$$= 338.14 \frac{\text{kJ}}{\text{kg}}.$$

Example 4.8

0.2 kg of CO_2 is contained in a rigid vessel of volume 0.1 m^3 and a pressure of 2 bar. Calculate the temperature of the gas in °C. Calculate also the energy added by heat transfer from an outside source if the temperature rises by 200 K. c_p = 1.0 kJ/(kg K).

Answer

For CO_2, $m_v = 12 + 32 = 44 \dfrac{\text{kg}}{\text{kmol}}$; also $R_0 = 8.3143 \dfrac{\text{kJ}}{\text{kmol K}}$ (p. 24 of tables).

Now
$$T = \frac{pV}{mR} = \frac{pVm_v}{mR_0} = \frac{200 \dfrac{\text{kN}}{\text{m}^2} \times 0.1 \text{ m}^3 \times 44 \dfrac{\text{kg}}{\text{kmol}}}{0.2 \text{ kg} \times 8.3143 \dfrac{\text{kg}}{\text{kmol K}}}$$

$$= 529.2 \text{ K} = 256.2 \text{ °C};$$

$$R = \frac{R_0}{m_v} = \frac{8.3143 \dfrac{\text{kJ}}{\text{kmol K}}}{44 \dfrac{\text{kg}}{\text{kmol}}} = 0.189 \frac{\text{kJ}}{\text{kg K}} \text{ for } CO_2;$$

$$c_v = c_p - R = 1.0 - 0.189 = 0.811 \frac{\text{kJ}}{\text{kg K}}.$$

Thus
$$E_2 - E_1 = m\, c_v\, (T_2 - T_1) = 0.2 \text{ kg} \times 0.811 \frac{\text{kJ}}{\text{kg K}} \times 200 \text{ K},$$

$$= 32.44 \text{ kJ}$$

for a non-flow process → this is a fixed mass of gas in a rigid vessel.

Exercises

1 Octane gas (C_8H_{18}) is heated in steady flow at constant pressure from 60 °C to 200 °C. The average c_p in this range may be taken as 2.22 kJ/(kg K). The energy added is 30 kW. Calculate the mass flow rate of gas. (0.0965 kg/s)

2 0.25 kg of air at 2.5 bar in a closed rigid vessel of volume 0.2 m^3 undergoes a positive heat transfer of 25 kJ. Calculate the initial and final temperatures. (Use p. 24 for the constants for air.) (423.9 °C, 563.2 °C)

As suggested in question 2, air is generally treated as a perfect gas at normal temperatures and the constants on p. 24 apply, namely
R = 0.287 kJ/(kg K); c_p = 1.005 kJ/(kg K); c_v = 0.718 kJ/(kg K).

3 A mass of oxygen occupies 0.5 m^3 at 2 bar and a temperature of 27 °C.
It is compressed according to the law $pV^{1.25}$ = constant until its volume is 0.2 m^3.
Calculate the heat and work transfers, assuming that c_v is 0.7 kJ/(kg K).
(Work *in* = –102.9 kJ, heat *out* = –33.57 kJ.)

5 First Law of Thermodynamics—Non-flow Processes

5.1 Introduction

The first law of thermodynamics is strictly *not* a law at all since it cannot be verified experimentally. It should be called the first axiom of thermodynamics since it cannot be disproved (rather a different statement).

What it really says is that energy cannot be created or destroyed, only transformed from one kind to another.

The simplest statement for non-flow processes is:

$$ {}_1Q_2 - {}_1W_2 = E_2 - E_1 $$

where the symbols are as usual.

Rewriting,

$$ {}_1Q_2 - {}_1W_2 = m(e_2 - e_1) $$

or $\quad {}_1q_2 - {}_1w_2 = e_2 - e_1 \quad$ per unit mass.

It must be clearly understood that the negative sign of the above equation is quite independent of the sign of either q or w which is decided by inspection of the problem and the chosen convention already given (i.e., q positive when added to the system, w negative when delivered to the system and vice versa).

In differential form we can write

$$ \delta q - \delta w = \mathrm{d}e. $$

(Note again the use of δ for q and w since they are not properties.)

For the special case of a reversible process

$$ \delta w_{\text{rev}} = p\,\mathrm{d}v $$

as already established.

(We see in Chapter 9, on the second law, that correspondingly for a reversible process $\delta q_{\text{rev}} = T\,\mathrm{d}s$.)

The correct use of the first law requires an examination of each term. For example, in a non-flow process involving a rigid vessel with no exterior connection for work transfer the law reduces to $\delta q = \mathrm{d}e$ (since $\delta w = 0$).

Again, for the adiabatic expansion of a mixed mass of fluid, $-\delta w = \mathrm{d}e$ (since adiabatic means $\delta q = 0$).

As mentioned earlier, the signs of q and w are crucial, e.g. w for a compression process is negative relative to the system and further ${}_1w_2 = -{}_2w_1$ by definition etc.

5.2 Worked Examples

Example 5.1

(a) A system consisting of a mixture of air and petrol at an initial temperature of 20 °C is contained in a rigid vessel. The mixture undergoes the following processes in sequence:

(i) Mixture temperature is raised to 250 °C by a positive heat transfer of 3 kJ.
(ii) Mixture ignites and burns completely and adiabatically and the temperature rises to 1500 °C.
(iii) The temperature of the products of combustion is reduced to 138 °C by a heat transfer out of –35 kJ. Evaluate the energy of the system after each process if the initial internal energy is 10 kJ.

(b) An equal mass of the *same mixture* is contained in a cylinder closed by a piston and undergoes the following processes in sequence.

(i) Adiabatic compression to a temperature of 250 °C.
(ii) Adiabatic combustion at constant volume until the burning is complete.
(iii) Expansion to a temperature of 138 °C, during which the work done by the system is 30 000 N m.
(iv) Cooling at constant volume until the temperature is again 20 °C.

On the assumption that the energy of the system depends only on its temperature and chemical aggregation (the way in which the elements are combined) evaluate the work done during (b) (i) and the heat transfer during (b) (ii). Did the system execute a cyclic process?

Answer

(a) For a rigid system throughout there can be no work done on the surroundings.

(i) ${}_1Q_2 - {}_1W_2 = E_2 - E_1$; or +3 kJ + 0 = E_2 – 10 kJ.
Thus E_2 = 13 kJ for the air–petrol mixture at 250 °C.
(ii) ${}_2Q_3 - {}_2W_3 = E_3 - E_2$; but for an adiabatic ${}_2Q_3 = 0$.
Thus 0 – 0 = $E_3 - E_2$ or $E_3 = E_2$ = 13 kJ at 1500 °C.
(iii) ${}_3Q_4 - {}_3W_4 = E_4 - E_3$ or –35 kJ – 0 = E_4 – 13 kJ
or E_4 = –22 kJ for combustion products at 138 °C.

(b) Same mixture.

(i) ${}_1Q_2 - {}_1W_2 = E_2 - E_1$; adiabatic compression and thus
E_2 = 13 kJ for same mixture at 250 °C.
(ii) ${}_2Q_3 - {}_2W_3 = E_3 - E_2$; adiabatic *and* constant volume
or 0 – 0 = $E_3 - E_2$ or $E_3 = E_2$ = 13 kJ and T = 1500 °C (as before).
(iii) ${}_3Q_4 = {}_3W_4 + E_4 - E_3$; E_4 for products at 138 °C is –22 kJ.
${}_3Q_4$ = 30 kJ – 22 kJ – 13 kJ = –5 kJ.
↑
${}_3W_4$
(iv) ${}_4Q_5 - {}_4W_5 = E_5 - E_4$; constant volume (${}_4W_5 = 0$).
System *cooled* to 20 °C → i.e. ${}_4Q_5$ is negative.
Hence $E_5 < E_4$ or $E_5 < -22$ kJ.

Thus initial and final states are *not* equal → the system has not executed a cycle (certainly not chemically).

Example 5.2

A mass of 2 kg of H_2O at 10 bar, 160 °C, is heated at constant pressure to 250 °C and then cooled at constant volume to 4 bar. Calculate the magnitudes and senses of the heat and work transfers in each of these processes.

Answer

The first thing to achieve is a clear *picture* of these events on one or more fields of state. In this case the same events are given on the T/v, p/v, T/h fields for the sake of clarity. But salient points have to be found before the diagrams can be drawn.

Now at 10 bar, $T_{sat} = 179.9$ °C. (p. 4 of tables)

But $T_1 = 160$ °C.

Thus the H_2O is *subcooled liquid* at 1.

Now at 10 bar, $T_2 = 250$ °C (i.e. $> T_{sat}$).

Thus the H_2O is *superheated vapour* at 2.

Further, $v_{10,250} = v_2 = v_3 = 0.2328$ m³/kg.

But v_g at 4 bar = 0.4623 m³/kg. (p. 4 of tables)

Thus H_2O is *wet vapour* at 3.

The sketches of the state diagrams may now be drawn, as here.

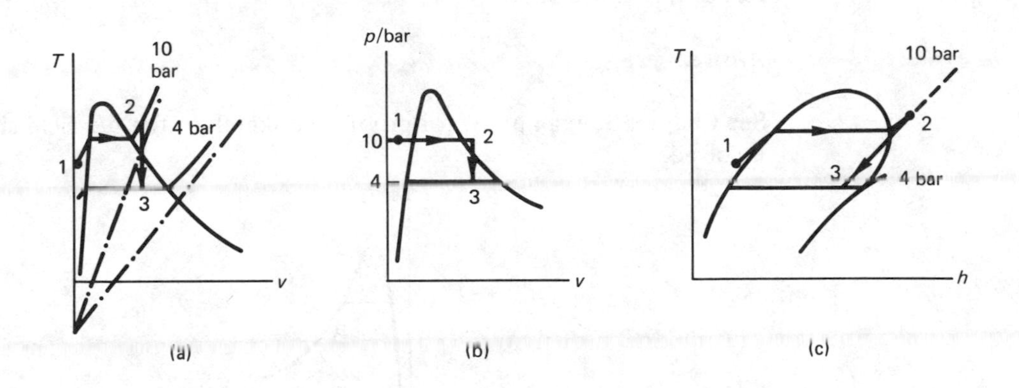

These diagrams are *not* to scale, as calculations will now show.

$v_1 = 0.001\,02$ m³/kg (p. 10); $v_2 = 0.2328$ m³/kg (already found).

Thus

$$_1w_2 = \int_{v_1}^{v_2} p\,dv \quad \text{and} \quad {}_1W_2 = m\int_{v_1}^{v_2} p\,dv = mp\,(v_2 - v_1 \ (p \text{ const.}),$$

$$\text{or} \quad {}_1W_2 = 2 \text{ kg} \times 1000 \frac{\text{kN}}{\text{m}^2} \times (0.2328 - 0.001\,102) \frac{\text{m}^3}{\text{kg}} = +463.4 \text{ kJ (delivered)}.$$

Now $$e_1 = 669 + \frac{1.2}{6.2}(27) = 674.2 \text{ kJ/kg}$$

(since 160 °C lies between 158.8 °C and 165 °C on p. 4).

Also, $e_2 = 2711$ kJ/kg. (p. 7)

Thus $_1Q_2 = {}_1W_2 + E_2 - E_1$ (1st law) $= {}_1W_2 + m\,(e_2 - e_1)$

$$= +\,463.4 \text{ kJ} + 2 \text{ kg } (2711 - 674.2) \text{ kJ/kg} = +4536.9 \text{ kJ (added)}.$$

$$_2W_3 = 0$$

(since volume is held constant here).

Also, $$x_3 \triangleq \frac{v_3}{v_{g,3}} \triangleq \frac{v_2}{v_{g,3}} \triangleq \frac{0.2328}{0.4623} = 0.502.$$

(This shows that point 3 *should* lie approximately halfway between f and g lines on all the diagrams!)

Now $e_3 = e_{f,3} + x_3(e_{g,3} - e_{f,3}) = 605 + 0.502\,(2554 - 605)$ kJ/kg (p. 4),

$e_3 = 1583.4$ kJ/kg.

Thus $_2Q_3 = {}_2W_3 + E_3 - E_2 = {}_2W_3 + m\,(e_3 - e_2)$,

$_2Q_3 = 0 + 2$ kg $(1583.4 - 2711)$ kJ/kg $= -\,2255.2$ kJ (rejected).

Example 5.3

Show that the work transfer per unit mass during a reversible non-flow process following the law pv^n = constant, from a state p_1, v_1 to a state p_2, v_2, is $(p_1v_1 - p_2v_2)/(n-1)$.

A mass of 1 kg of saturated steam at a pressure of 10 bar is expanded as above in a cylinder to 1 bar and a dryness of 0.88. Determine the work and heat transfers during the process.

Answer

See Chapter 3, example 1, for proof. The sketch of the p/v field shows the process involved.

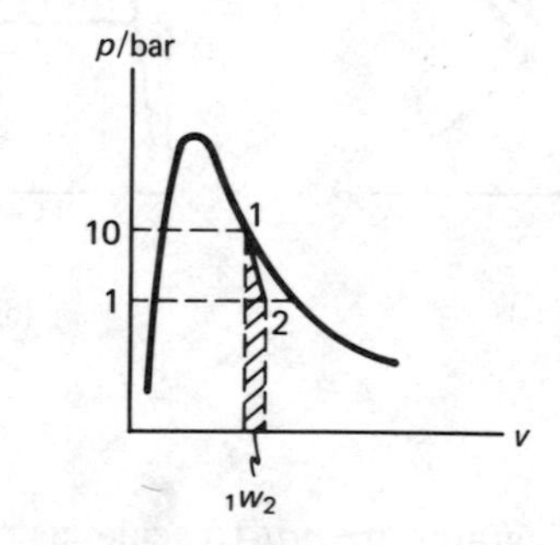

$$v_1 = v_{g,1} = 0.1944 \; \frac{\text{m}^3}{\text{kg}}. \text{ (p. 4)}$$

$$v_2 \triangleq x_2 v_{g,2} = 0.88\,(1.694) = 1.4907 \; \frac{\text{m}^3}{\text{kg}}. \text{ (p. 4)}$$

Now $p_1v_1^n = p_2v_2^n$, or, taking natural logs on both sides,

$$\ln p_1 + n \ln v_1 = \ln p_2 + n \ln v_2.$$

Thus

$$n = \frac{\ln(p_1/p_2)}{\ln(v_2/v_1)} = \frac{\ln\left(\dfrac{10}{1}\right)}{\ln\left(\dfrac{1.4907}{0.1944}\right)} = \frac{2.3026}{2.037} = 1.13.$$

Thus

$${}_1w_2 = \frac{p_1v_1 - p_2v_2}{n-1} = \frac{\left(1000\ \frac{\text{kN}}{\text{m}^2} \times 0.1944\ \frac{\text{m}^3}{\text{kg}}\right) - \left(10\ \frac{\text{kN}}{\text{m}^2} \times 1.4907\ \frac{\text{m}^3}{\text{kg}}\right)}{0.13},$$

$${}_1w_2 = +1380.7\ \frac{\text{kJ}}{\text{kg}}\ \text{(delivered)}.$$

Now $e_1 = e_{g,1} = 2584$ kJ/kg. (p. 4)

And $e_2 = e_{f,2} + x_2\,(e_{g,2} - e_{f,2}) = 417 + 0.88\,(2506 - 417)$,

$= 2255.32$ kJ/kg.

Thus ${}_1q_2 = {}_1w_2 + e_2 - e_1 = 1380.7 + 2255.32 - 2584$

$= +1052.02$ kJ/kg (added).

Example 5.4

(a) Consider the four strokes of a reciprocating i.c. engine:

(i) *Suction:* suction valve open – charge induced as piston moves down cylinder to bottom dead centre.
(ii) *Compression:* both valves shut – charge compressed into clearance space.
(iii) *Expansion:* charge expands to exhaust pressure.
(iv) *Exhaust:* exhaust valve open – most of the charge is exhausted as the piston moves to top dead centre.

Select, for all of the above strokes (i)–(iv), the appropriate form of the the energy equation from those listed below:

(A) $E_1 + {}_1Q_2 = E_2 + {}_1W_2$;
(B) $E_1 + H_T + {}_1Q_2 = E_2 + {}_1W_2$;
(C) $H_1 + {}_1Q_2 = H_2 + {}_1W_2$;
(D) $H_1 + H_T + {}_1Q_2 = H_2 + {}_1W_2$;

where E = internal energy, H = enthalpy, subscript 1 = state at the start of any stroke, subscript 2 = state at the end of any stroke, and subscript T = state of any transfer charge.

(b) Steam at 6 bar, 250 °C, is expanded to 1 bar according to the law pv^n = constant. Calculate the specific work transfer in kJ/kg.

Answer

(a) (i) Consider the start, middle and finish of the suction process by reference to the diagrams here.

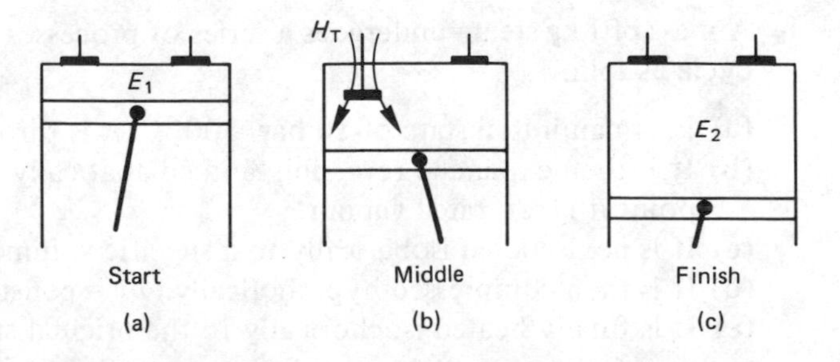

The appropriate energy equation here is (B) since the clearance mass is a fixed packet of fluid at the start, (before the suction valve opens) and possesses internal energy E_1. The fresh charge clearly has enthalpy H_T since it *flows* into the cylinder, but once the suction valve shuts again at the end of the stroke the total mass in the cylinder (clearance plus fresh charge) now possesses internal energy E_2. Since the piston has moved there is a work transfer to the surroundings and there may well be a small but finite heat transfer (although this is often ignored in elementary calculations).

(ii) The appropriate energy equation is now (A) since the energy in the cylinder at the start of the compression process is identical to that at the end of the previous process. Thus E_2 in the previous process now becomes E_1 in this process. There is no transfer term this time since no valve opens in compression.

(iii) The energy equation is again (A) since again no valve opens in expansion.

(iv) The energy equation is now (B) again since this time the exhaust valve opens to allow the exhuast products to escape from the cylinder.
(Note that strictly suction and exhaust are not non-flow processes but unsteady-flow processes since mass crosses the system boundary during them.)

(b) $$v_1 = 0.394 \frac{\text{m}^3}{\text{kg}}\ ;\ \text{(p. 7 of tables)}$$

$$p_1 v_1^n = p_2 v_2^n$$

or $$v_2^n = \frac{p_1}{p_2} \times v_1^n = \frac{6}{1}(0.394)^{1.3} = 1.7877,$$

$$v_2 = 1.7877^{1/1.3} = 1.7877^{0.7692} = 1.5634 \frac{\text{m}^3}{\text{kg}}.$$

$${}_1w_2 = \int_{v_1}^{v_2} p\,\mathrm{d}v = \frac{p_1 v_1 - p_2 v_2}{n-1} \quad \text{(see previous work)}$$

$$= \frac{\left(600 \frac{\text{kN}}{\text{m}^2} \times 0.394 \frac{\text{m}^3}{\text{kg}}\right) - \left(100 \frac{\text{kN}}{\text{m}^2} \times 1.5634 \frac{\text{m}^3}{\text{kg}}\right)}{0.3}$$

$$= 266.87 \frac{\text{kJ}}{\text{kg}}.$$

***Example** 5.5*

A mass of 1 kg steam undergoes a series of processes constituting a thermodynamic cycle as follows:

(a) From an initial state of 10 bar, 500 °C, it is throttled to 4 bar.
(b) It is then expanded reversibly and adiabatically to a pressure of 1 bar at which point it is saturated vapour.
(c) It is next cooled isobarically to a specific volume of 0.6 m^3/kg.
(d) It is then compressed hyperbolically (pv = const) to its original volume.
(e) It is finally heated isochorically to the original state at 10 bar.

Calculate the work and heat transfers in each of the above processes and draw clear state diagrams to illustrate the cycle.

Answer

(a) In a throttling process $h_1 = h_2$. (See standard texts.) The implication here is that ${}_1w_2 = {}_1q_2 = 0$.

Now $h_1 = h_2 = 3478 \dfrac{\text{kJ}}{\text{kg}}$, (p. 7 of tables, at 10 bar, 500 °C)

$$v_1 = 0.354 \frac{\text{m}^3}{\text{kg}}.$$

Thus $T_2 = 400\,°\text{C} + \left(\dfrac{3478 - 3274}{3485 - 3274}\right)(500 - 400)\ \text{K}$ (interpolating on p. 6 at 4 bar)

or $T_2 = 496.7\,°\text{C}$.

Thus $v_2 = 0.7725 + 0.967\,(0.8893 - 0.7725) = 0.8854 \dfrac{\text{m}^3}{\text{kg}}$ (interpolating on p. 6)

and $e_2 = 2965 + 0.967\,(3129 - 2965) = 3123.6 \dfrac{\text{kJ}}{\text{kg}}$. (interpolating on p. 6)

(b) The process is adiabatic and thus

$${}_2q_3 = 0.$$

The fluid is saturated vapour at 3.

Thus $v_3 = v_{g,3} = 1.694 \dfrac{\text{m}^3}{\text{kg}}$ at 1 bar.

also $e_3 = e_{g,3} = 2506 \dfrac{\text{kJ}}{\text{kg}}$

and $-{}_2w_3 = e_3 - e_2$

or ${}_2w_3 = e_2 - e_3 = 3123.6 - 2506 = +617.6 \dfrac{\text{kJ}}{\text{kg}}$.

(c) $x_4 = \dfrac{v_4}{v_{g,4}} = \dfrac{v_4}{v_{g,3}}$ (since $p_3 = p_4$)

$$= \frac{0.6}{1.694} = 0.354.$$

$$e_4 = e_f + x\,(e_g - e_f) = 417 + 0.354\,(2506 - 417) = 1156.5 \frac{\text{kJ}}{\text{kg}}.$$

$${}_3w_4 = \int_{v_3}^{v_4} p\,\mathrm{d}v = p_3(v_4 - v_3) = 100 \frac{\text{kN}}{\text{m}^2}\,(0.6 - 1.694)\,\frac{\text{m}^3}{\text{kg}} = -109.4 \frac{\text{kJ}}{\text{kg}}.$$

$${}_3q_4 = {}_3w_4 + e_4 - e_3 = -109.4 \frac{\text{kJ}}{\text{kg}} + 1156.5 \frac{\text{kJ}}{\text{kg}} - 2506 \frac{\text{kJ}}{\text{kg}};$$

$$= -1458.9 \frac{\text{kJ}}{\text{kg}}.$$

(d) $p_4v_4 = p_5v_5 = p_6v_1$ (since $v_5 = v_1$).

Thus $p_5 = p_4 \times \dfrac{v_4}{v_1} = 1\ \text{bar} \times \dfrac{0.6}{0.354} = 1.695\ \text{bar}.$

Now at 1.695 bar,

$$v_{g,5} = 1.091 + 0.95\,(1.031 - 1.091) = 1.034 \frac{\text{m}^3}{\text{kg}}. \quad \text{(p. 4)}$$

Thus $$x_5 = \frac{v_5}{v_{g,5}} = \frac{v_1}{v_{g,5}} = \frac{0.354}{1.034} = 0.342.$$

Now a double interpolation (vertical + horizontal) is required.

$$e_{1.6,\, x=0.342} = 475 + 0.342\,(2521 - 475) = 1174.7\ \frac{\text{kJ}}{\text{kg}}. \qquad \text{(p. 4)}$$

$$e_{1.7,\, x=0.342} = 483 + 0.342\,(2524 - 483) = 1181\ \frac{\text{kJ}}{\text{kg}}.$$

Thus $$e_{1.695,\, x=0.342} = 1174.7 + 0.95\,(1181 - 1174.7) = 1180.7\ \frac{\text{kJ}}{\text{kg}} = e_5.$$

And $$_4w_5 = \int_{v_4}^{v_5} p\,\mathrm{d}v = \int_{v_4}^{v_5} \frac{\mathrm{K}\,\mathrm{d}v}{v} = p_4 v_4 \ln \frac{v_5}{v_4}$$

$$= 100\ \frac{\text{kN}}{\text{m}^2} \times 0.6\ \frac{\text{m}^3}{\text{kg}} \ln \frac{0.342}{0.6} = -33.7\ \frac{\text{kJ}}{\text{kg}}.$$

And $$_4q_5 = {_4w_5} + e_5 - e_4 = -33.7 + 1180.7 - 1156.5 = -9.5\ \frac{\text{kJ}}{\text{kg}}.$$

(e) $$v_6 = v_5 = v_1 = 0.354\ \frac{\text{m}^3}{\text{kg}};$$

$$p_6 = p_1 = 10\ \text{bar} \quad \text{(original state)};$$

$$_5w_6 = 0 \quad \text{(constant volume)};$$

$$e_6 = e_1 = 3124\ \frac{\text{kJ}}{\text{kg}}.$$

Thus $$_5q_6 = {_5w_6} + e_6 - e_5 = 0 + 3124 - 1180.7$$

$$= 1943.3\ \frac{\text{kJ}}{\text{kg}}.$$

A check will soon show that $h_5 > h_4$ and 5 lies to right of 4 in diagram (b).

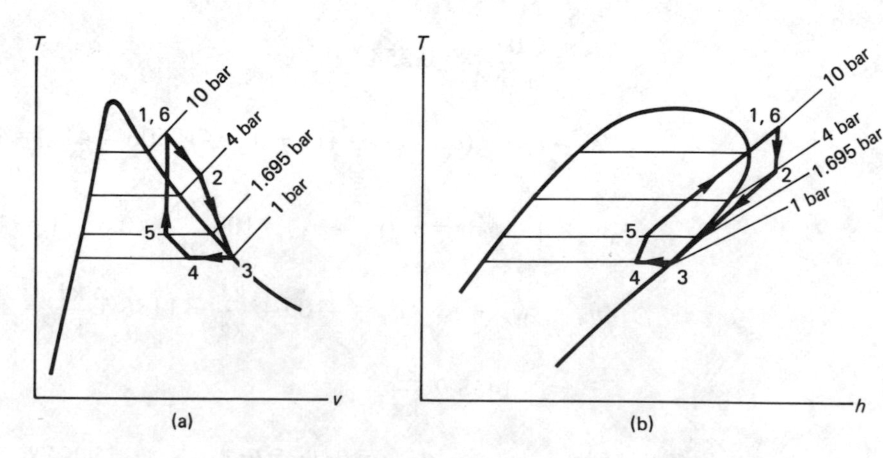

5.3 Perfect Gases

Further consequences of the properties of a perfect gas for a reversible adiabatic process are as follows.

First law in differential form is

$$\delta q - \delta w = \mathrm{d}e$$

and for an adiabatic $\delta w = -\,\mathrm{d}e$ since $\delta q = 0$;

also, for a gas, $\delta w = -\,c_v\,\mathrm{d}T = \dfrac{-R}{\gamma - 1}\,\mathrm{d}T$

(since $c_p/c_v = \gamma$ and $c_p - c_v = R$).

Thus $${}_1w_2 = -\int_{T_1}^{T_2} \frac{R}{\gamma - 1}\,\mathrm{d}T = \frac{R}{\gamma - 1}(T_1 - T_2).$$

But $${}_1w_2 = \frac{R}{n - 1}(T_1 - T_2) \qquad \text{when } pv^n = \mathrm{K}.$$

Thus $pv^\gamma = \mathrm{K}$ for a reversible adiabatic (i.e. $n = \gamma$).

Furthermore, $$\frac{p_1 v_1}{T_1} = \frac{p_2 v_2}{T_2}$$

and if $pv^n = \mathrm{K}$ then $p_1 v_1^n = p_2 v_2^n$

and on eliminating v between these we get

$$\frac{T_2}{T_1} = \left(\frac{p_1}{p_1}\right)^{(n-1)/n}$$

and eliminating p,

$$\frac{T_2}{T_1} = \left(\frac{v_1}{v_2}\right)^{n-1}$$

for a reversible polytropic process.

Thus for a reversible adiabatic substitute $\gamma = n$ in these expressions.

5.4 Worked Examples

Example 5.6

Air is compressed reversibly in a cylinder according to the process equation $pv^{1.3}$ = constant. The air is initially at a pressure and temperature of 1 bar, 300 K and the final pressure is 6 bar.

Calculate the heat transfer between the air and its surroundings per unit mass during the process and state its sense.

What would have been the final pressure had the air been compressed through the same volume ratio adiabatically?

Answer

$$T_2 = T_1\left(\frac{p_2}{p_1}\right)^{(n-1)/n} = 300\ \mathrm{K} \times 6^{0.3/1.3} = 453.62\ \mathrm{K}.$$

$$ {}_1w_2 = \frac{R(T_1 - T_2)}{n-1} = \frac{0.287 \dfrac{\text{kJ}}{\text{kg K}} (300 - 453.62)\ \text{K}}{0.3} = -146.97 \frac{\text{kJ}}{\text{kg}} $$

$$ {}_1q_2 = {}_1w_2 + e_2 - e_1 = {}_1w_2 + c_v (T_2 - T_1) $$

$$ = -146.97 \frac{\text{kJ}}{\text{kg}} + 0.718 \frac{\text{kJ}}{\text{kg K}} (453.62 - 300)\ \text{K} = -146.97 + 110.3 $$

$$ {}_1q_2 = -36.67 \frac{\text{kJ}}{\text{kg}} \qquad \text{(rejected by the gas).} $$

For a reversible adiabatic process pv^γ = constant.

Now $$ \frac{v_1}{v_2} = \left(\frac{p_2}{p_1}\right)^{1/n} = 6^{1/1.3} = 3.9681. $$

Thus $$ \frac{p_2}{p_1} = \left(\frac{v_1}{v_2}\right)^{\gamma} = 3.9681^{1.4} = 6.887 \qquad (\gamma = 1.4 \text{ on p. 24}) $$

and $$ p_2 = 6.887 \text{ bar.} $$

***Example** 5.7*

A mass of 1 kg of a perfect gas at a pressure of 1 bar and a temperature of 15 °C when compressed reversibly according to the law $pv^{1.2}$ = constant requires a work transfer of 304.22 kJ and the compression is accompanied by a heat transfer of 152.11 kJ from the gas. The final temperature is 227 °C.

What would be the work transfer required to compress the gas through the same pressure ratio by a reversible adiabatic process and what would be the final temperature?

Answer

$$ {}_1w_2 = \frac{R(T_1 - T_2)}{n} \quad \text{per kg gas and is negative in compression.} $$

Thus $$ -304.22 \frac{\text{kJ}}{\text{kg}} = \frac{R \times (15 - 227)\ \text{K}}{0.2} $$

or $$ R = \frac{304.22 \times 0.2}{212} = 0.287 \frac{\text{kJ}}{\text{kg K}}. $$

(Note that T_1, T_2 are quoted here in Celsius since in subtraction the necessary addition of 273 to each cancels out.)

$$ {}_1q_2 - {}_1w_2 = e_2 - e_1 = c_v (T_2 - T_1). $$

Thus $$ c_v = \frac{{}_1q_2 - {}_1w_2}{T_2 - T_1} = \frac{-152.11 \dfrac{\text{kJ}}{\text{kg}} - \left(-304.22 \dfrac{\text{kJ}}{\text{kg}}\right)}{212\ \text{K}} = 0.7175 \frac{\text{kJ}}{\text{kg K}}. $$

(*Note:* ${}_1q_2$ is also negative.)

Now $$ R = c_p - c_v $$

and $$ \frac{R}{c_v} = \frac{c_p}{c_v} - 1 $$

or $$ \gamma = \frac{R}{c_v} + 1 = \frac{0.287 \dfrac{\text{kJ}}{\text{kg K}}}{0.7175 \dfrac{\text{kJ}}{\text{kg K}}} + 1 = 1.4. $$

$$\frac{p_2}{p_1} = \left(\frac{T_2}{T_1}\right)^{n/(n-1)} = \left(\frac{500}{288}\right)^{1.2/0.2} \quad (T \textit{ must} \text{ be absolute here!})$$

$$\frac{p_2}{p_1} = 27.382.$$

For a reversible adiabatic compression.

$$\frac{T_2}{T_1} = \left(\frac{p_2}{p_1}\right)^{(\gamma-1)/\gamma}$$

or

$$T_2 = 288 \text{ K } (27.382)^{0.4/1.4} = 741.48 \text{ K } (468.48\ ^\circ\text{C}).$$

Work transfer required is given by

$${}_1W_2 = \frac{R\,(T_1 - T_2)}{\gamma - 1} = \frac{0.287 \dfrac{\text{kJ}}{\text{kg K}}\ (15 - 468.48) \text{ K}}{0.4} = -325.37 \text{ kJ}.$$

Example 5.8

A mass of 1 kg of air initially at 5 bar and 100 °C undergoes a cycle consisting of the following processes:

(a) Constant-pressure expansion until the volume is doubled.
(b) Constant-volume cooling.
(c) Reversible adiabatic compression to the initial state.

Calculate the pressure and temperature after the constant-volume cooling process and the net work done in the cycle.

Answer

The cycle can be shown on a sketch of the gaseous region of the p–v field, as here.

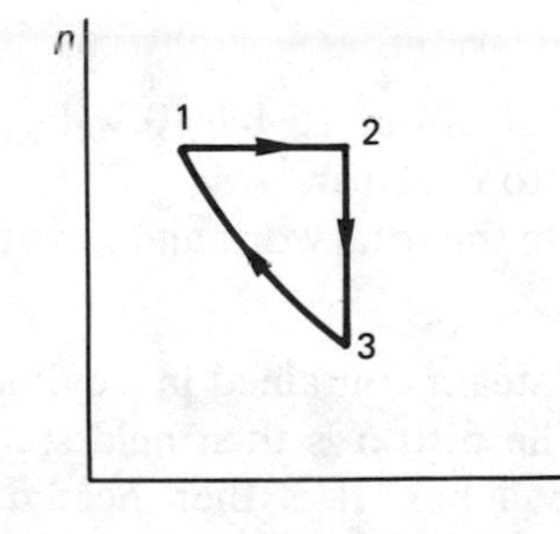

For a perfect gas at constant pressure using the gas laws

$$\frac{v_2}{T_2} = \frac{v_1}{T_1}$$

or

$$T_2 = T_1 \frac{v_2}{v_1} = 373 \text{ K} \times 2 = 746 \text{ K}.$$

For constant volume

$$\frac{T_2}{T_3} = \left(\frac{p_2}{p_3}\right),$$

and for the reversible adiabatic process

$$\frac{T_1}{T_3} = \left(\frac{p_1}{p_3}\right)^{(\gamma-1)/\gamma} = \left(\frac{p_2}{p_3}\right)^{(\gamma-1)/\gamma} = \left(\frac{T_2}{T_3}\right)^{(\gamma-1)/\gamma}$$

using the previous expression.

$$\text{Thus} \quad \frac{T_1}{T_2^{(\gamma-1)/\gamma}} = \frac{T_3}{T_3^{(\gamma-1)/\gamma}} = T_3^{1/\gamma}$$

$$\text{or} \quad T_3 = \frac{T_1^{\gamma}}{T_2^{\gamma-1}} = \frac{373^{1.4}}{746^{0.4}} = 282.68 \text{ K.}$$

$$p_3 = p_1 \left(\frac{T_3}{T_1}\right)^{\gamma/(\gamma-1)} = 5 \text{ bar} \left(\frac{282.68}{373}\right)^{1.4/0.4} = 1.8946 \text{ bar.}$$

$$\text{Now} \quad v_1 = \frac{RT_1}{p_1} = \frac{0.287 \dfrac{\text{kJ}}{\text{kg K}} \times 373 \text{ K}}{500 \dfrac{\text{kN}}{\text{m}^2}} \left[\frac{\text{kN m}}{\text{kJ}}\right] = 0.2141 \frac{\text{m}^3}{\text{kg}}.$$

$$\text{Thus} \quad v_2 = 0.4282 \frac{\text{m}^3}{\text{kg}};$$

$${}_1w_2 = p_1 (v_2 - v_1) = 500 \frac{\text{kN}}{\text{m}^2} (0.4282 - 0.2141) \frac{\text{m}^3}{\text{kg}} = +107.05 \frac{\text{kJ}}{\text{kg}};$$

$${}_2w_3 = 0 \quad (v \text{ is constant});$$

$${}_3w_1 = \frac{R(T_3 - T_1)}{\gamma - 1} = \frac{0.287 \dfrac{\text{kJ}}{\text{kg K}} (282.68 - 373) \text{ K}}{0.4} = -64.8 \frac{\text{kJ}}{\text{kg}};$$

$$\text{and net work} = 107.05 - 64.8 = 42.25 \frac{\text{kJ}}{\text{kg}}.$$

Exercises

1 1 kg of steam at 15 bar, 0.9 dry, is throttled to 0.05 bar and then cooled isochorically to 0.01 bar.

Calculate the total work and heat transfers for the two processes taken together.

(Net work = 0, net heat = –2044.96 kJ/kg.)

2 1 kg of steam contained in a cylinder at 10 bar, 150 °C, is heated isobarically to 250 °C. The piston is then held stationary while the steam is cooled at constant volume to 4 bar. It is then heated again at constant pressure (with the piston released again) to 200 °C.

Calculate the total work and heat transfers involved.

(Work transfer = 352.38 kJ/kg, heat = 2447.78 kJ/kg.)

3 1 kg of steam at 10 bar, 0.4 dry, is heated in a cylinder to saturated vapour at constant pressure. The piston is then held fixed and isochoric cooling follows to a pressure of 2 bar.

Calculate the value of n, the polytropic index necessary for a final polytropic process (pv^n = constant) which will restore the steam to its initial state, and find the net cyclic work and heat transfers.

(n = 1.757, cyclic work = 65.2 kJ/kg = cyclic heat.)

4 A perfect gas for which $\gamma = 1.4$ undergoes a thermodynamic cycle consisting of three processes:

(a) Isothermal compression from an initial temperature of 25 °C to $\frac{1}{7}$ of its original volume.
(b) *Either* constant volume addition of energy
or constant pressure addition of energy.
(c) Reversible adiabatic expansion to the initial state.

Sketch these alternative cycles on a p–v field and calculate the maximum temperature for each case.

Find also the ratio of net work to energy added for each cycle.

(649 K, 519.9 K, 0.339, 0.253)

6 First Law of Thermodynamics—Flow Processes

6.1 Introduction

This chapter is confined to the application of the first law of thermodynamics to *steady*-flow processes, i.e. those processes defined by two fundamental principles:

(a) Steady mass flow rate $\dot{m} = \rho A u$ (mass continuity)
where ρ = fluid density
A = duct area
u = fluid velocity
and $\dot{m}$ = steady mass flow rate $= \left(\frac{\mathrm{d}m}{\mathrm{d}t}\right)$.
(This is usually recast as $\dot{m}v = uA$ since $v = 1/\rho$.)

(b) Conservation of energy (the steady-flow energy equation or SFEE for short):

$$_1\dot{Q}_2 - {}_1\dot{W}_2 = \dot{m}\,[(h_2 - h_1) + \tfrac{1}{2}(u_2^2 - u_1^2) + g(z_2 - z_1)]$$

where z = height above datum (all other symbols as before).

Note that very often gz is negligible compared with other terms and the reduced SFEE is written as

$$_1\dot{Q}_2 - {}_1\dot{W}_2 = \dot{m}\,[(h_2 - h_1) + \tfrac{1}{2}(u_2^2 - u_1^2)] \qquad \text{(per unit mass) flow rate.}$$

$$_1q_2 - {}_1w_2 = (h_2 - h_1) + \tfrac{1}{2}(u_2^2 - u_1^2)$$

An even further reduction is given by

$$_1q_2 - {}_1w_2 = h_2 - h_1 \qquad \text{(for negligible k.e. terms).}$$

Many processes in reality approximate closely to steady flow, e.g. steady conditions in a steam power plant after the 'start up' transient is over and steady conditions exist at all points in the system.

Note again the two essential characteristics of steady flow:

(a) $\dot{m}$ is constant,
(b) the properties at any station are invariable with time (they may of course vary from station to station).

The one obvious application of the SFEE where gz is prominent is in a hydroelectric plant where a change in z is essential for power production.

6.2 Worked Examples

Example 6.1

Steam is supplied to a turbine at the rate of 0.5×10^6 kg/h. The state of the steam entering the turbine is 150 bar, 500 °C, and it enters with negligible velocity. The exhaust steam leaves the turbine at 0.15 bar, 0.9 dry via a passage of cross-sectional area 10 m².

Determine:

(a) the specific volume and velocity of the exhaust steam,
(b) the power output and heat rate from the turbine if the heat transfer is 0.05 per cent of the power output.

$$v_2 = \frac{(pv)_{0.14\ \text{bar}} + (pv)_{0.16\ \text{bar}}}{2 \times 0.15\ \text{bar}} \quad \text{(care needed here at low pressure!)}$$

$$= \left\{ 14\ \frac{\text{kN}}{\text{m}^2}\ [(0.1 \times 0.001\,014) + (0.9 \times 10.69)]\ \frac{\text{m}^3}{\text{kg}} \right.$$

$$\left. +16\ \frac{\text{kN}}{\text{m}^2}\ [(0.1 \times 0.001\,016) + (0.9 \times 9.432)]\ \frac{\text{m}^3}{\text{kg}} \right\} \Big/ 2 \times 15\ \frac{\text{kN}}{\text{m}^2}.$$

(*Note:* this is a *full* calculation of v for wet steam given by

$v = (1 - x)\, v_f + x v_g$ (v_f values on p. 10 of tables; v_g values on p. 3).)

$$v_2 = 9.0173\ \frac{\text{m}^3}{\text{kg}}.$$

Mass continuity:

$$u_2 = \frac{\dot{m} v_2}{A} = \frac{0.5 \times 10^6\ \dfrac{\text{kg}}{\text{h}} \left[\dfrac{\text{h}}{3600\ \text{s}}\right] \times 9.0173\ \dfrac{\text{m}^3}{\text{kg}}}{10\ \text{m}^2} = 125.2\ \frac{\text{m}}{\text{s}}.$$

Now $$h_2 = \frac{[220 + 0.9\,(2376)] + [232 + 0.9\,(2369)]}{2} = 2361.3\ \frac{\text{kJ}}{\text{kg}}$$

(interpolating for $h_f + x h_{fg}$ on p.3)

and $$h_1 = 3309\ \frac{\text{kJ}}{\text{kg}} \quad \text{(p. 8)}.$$

Energy equation:

$${}_1\dot{Q}_2 - {}_1\dot{W}_2 = \dot{m}\,[(h_2 - h_1) + \tfrac{1}{2} u_2^2] \qquad (u_1 = 0);$$

Thus $$-0.0005\ {}_1\dot{W}_2 - {}_1\dot{W}_2 = \dot{m}\,(h_2 - h_1) + \tfrac{1}{2} u_2^2 \qquad ({}_1\dot{Q}_2 \text{ is } -\text{ve})$$

or $$-1.005\ {}_1\dot{W}_2 = 500\,000\ \frac{\text{kg}}{\text{h}} \left[\frac{\text{h}}{3600\ \text{s}}\right] \left\{ (2361.3 - 3309)\ \frac{\text{kJ}}{\text{kg}} \right.$$

$$\left. + 125.2^2\ \frac{\text{m}^2}{\text{s}^2} \left[\frac{\text{N s}^2}{\text{kg m}}\right] \left[\frac{\text{kN}}{10^3\ \text{N}}\right] \right\} \left[\frac{\text{MW}}{10^3\ \text{kW}}\right]$$

or $${}_1\dot{W}_2 = 129.4\ \text{MW (delivered)}$$

and $${}_1\dot{Q}_2 = -0.0005\,(129.4)\ \text{MW} = -0.0647\ \text{MW} \quad \text{(64.7 kW OUT)}.$$

Example 6.2

(a) Define briefly and clearly the meaning of the term *steady flow*. Define enthalpy and explain clearly the difference between enthalpy and internal energy.

(b) Freon-12 is compressed from saturated vapour, 1.004 bar to 4.914 bar, 45 °C, in steady flow.

The inlet pipe has a diameter of 0.15 m and the flow velocity therein is 30 m/s.

Assuming that the heat transfer rate from the compressor is 5 per cent of the power output and that kinetic and potential energy changes are small, calculate the power input.

(c) Calculate the power input necessary for the same compressor with the same ratio of heat to work transfer as in (b) if the fluid were changed to air (assumed a perfect gas) between the same overall temperature limits as in (b) and with the same inlet duct diameter and velocity of flow.

Answer

(a) Refer to previous definitions in this chapter for steady flow and to Chapter 2 for clarification of e and h.

(b) *Mass continuity:*

$$\dot{m} = \frac{u_1 A_1}{v_1} = \frac{30 \dfrac{\text{m}}{\text{s}} \times \dfrac{\pi \times 0.15^2}{4} \text{m}^2}{0.1594 \dfrac{\text{m}^3}{\text{kg}}} = 3.326 \frac{\text{kg}}{\text{s}}$$

($v_1 = v_g$ at 1.004 bar, p. 13 of tables).

Steady-flow energy equation:

$${}_1\dot{Q}_2 - {}_1\dot{W}_2 = \dot{m}\,(h_2 - h_1) \qquad \text{(all other terms zero).}$$

Substituting for ${}_1\dot{Q}_2$,

$$(-0.05\ {}_1\dot{W}_2) - {}_1\dot{W}_2 = \dot{m}\,(h_2 - h_1).$$

Thus $${}_1\dot{W}_2 = \frac{\dot{m}\,(h_1 - h_2)}{1.05} = \frac{3.326 \dfrac{\text{kg}}{\text{s}} (174.2 - 214.35) \dfrac{\text{kJ}}{\text{kg}}}{1.05} \qquad \text{(p. 13)}$$

$$= -127.2 \text{ kW} \qquad \text{(i.e. IN).}$$

(c) For a gas

$$T_1 = -30 + 273 = 243 \text{ K} \qquad (T_{sat} = -30\ °\text{C at 1.004 bar})$$

$$T_2 = 273 + 45 = 318 \text{ K} \qquad \text{(at 45 °C).}$$

Thus $$v_1 = \frac{RT_1}{p_1} = \frac{0.287 \dfrac{\text{kJ}}{\text{kg K}} \times 243 \text{ K}}{100.4 \dfrac{\text{kN}}{\text{m}^2}} = 0.695 \frac{\text{m}^3}{\text{kg}}.$$

And $$\dot{m} = \frac{A_1 u_1}{v_1} = \pi \times \frac{0.15^2}{4} \text{m}^2 \times 30 \frac{\text{m}}{\text{s}} \times \frac{1}{0.695} \frac{\text{kg}}{\text{m}^3} = 0.763 \frac{\text{kg}}{\text{s}}.$$

$$\text{Thus} \quad {}_1\dot{W}_2 = \frac{\dot{m}\,(h_1 - h_2)}{1.05} \qquad \text{(as before)} \qquad = \frac{\dot{m}\,c_p\,(T_1 - T_2)}{1.05}$$

$$= 0.763\ \frac{\text{kg}}{\text{s}} \times 1.005\ \frac{\text{kJ}}{\text{kg K}}\ (-75\ \text{K}) = -54.8\ \text{kW} \qquad \text{(i.e. IN).}$$

Example 6.3

Steam is supplied to a turbine at a pressure of 150 bar and a temperature of 500 °C through a pipe 0.25 m in diameter. The steam is exhausted at 0.1 bar with a dryness fraction of 0.95 through a duct of 5 m² cross-sectional area. When supplied with steam at the rate of 25 kg/s the turbine developes 20 MW of power output.

Neglecting potential energy changes, calculate the rate of heat transfer from the turbine.

Answer

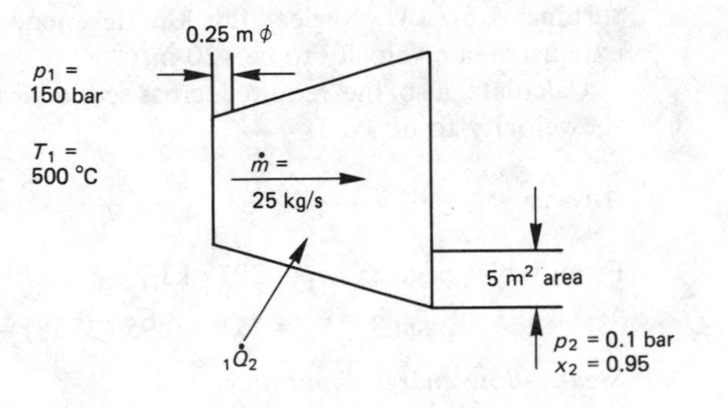

From tables page 8: $h_1 = 3309\ \dfrac{\text{kJ}}{\text{kg}}$.

page 3: $h_2 = h_{f,2} + x_2 h_{fg,2}$

or $h_2 = 192 + 0.95\,(2392) = 2464.4\ \dfrac{\text{kJ}}{\text{kg}}$.

Also, $v_1 = 0.020\,78\ \dfrac{\text{m}^3}{\text{kg}}$; (p. 8)

$$v_2 = x_2 v_{g,2} + (1 - x_2)\,v_{f,2} = 0.95\,(14.76) + 0.05\,(0.001) = 14.02\ \frac{\text{m}^3}{\text{kg}}.$$

Mass continuity:

$$u_1 = \frac{\dot{m}v_1}{A_1} = \frac{25\ \frac{\text{kg}}{\text{s}} \times 0.020\,78\ \frac{\text{m}^3}{\text{kg}}}{\frac{\pi}{4} \times (0.25\ \text{m})^2} = 10.583\ \frac{\text{m}}{\text{s}}$$

and

$$u_2 = \frac{25 \times 14.02}{5} = 70.11\ \frac{\text{m}}{\text{s}}.$$

Steady-flow energy equation:

$${}_1\dot{Q}_2 - {}_1\dot{W}_2 = \dot{m}\left[(h_2 - h_1) + \left(\frac{u_2^2 - u_1^2}{2}\right)\right].$$

Thus $\quad {}_1\dot{Q}_2 - 20\ \text{MW} = 25\ \frac{\text{kg}}{\text{s}} \left\{ (2464.4 - 3309)\ \frac{\text{kJ}}{\text{kg}} + \frac{\left(70.11\ \frac{\text{m}}{\text{s}}\right)^2 - \left(10.583\ \frac{\text{m}}{\text{s}}\right)^2}{2} \left[\frac{\text{kJ}}{10^3\ \text{N m}}\right] \left[\frac{\text{N s}^2}{\text{kg m}}\right] \right\}$

or $\quad {}_1\dot{Q}_2 - 20\ \text{MW} = 25 \left\{ \frac{\text{kg}}{\text{s}}\ (-844.6 + 2.402)\ \frac{\text{kJ}}{\text{kg}} \left[\frac{\text{MW}}{10^3\ \text{kW}}\right] \right\}.$

Thus $\quad {}_1\dot{Q}_2 = -21.055\ \text{MW} + 20\ \text{MW} = -1.055\ \text{MW}.$

Note that the arbitrarily drawn sense of ${}_1\dot{Q}_2$ in the sketch is in error but the sign of ${}_1\dot{Q}_2$ takes care of itself!

Example 6.4

Calculate the required mass flow rate for a steam turbine having a power output of 5 MW when it is supplied with steam at 150 bar and 400 °C if the steam exhaust conditions are 0.2 bar and 0.95 dry. The heat transfer rate from the turbine is 50 kW. Neglect the kinetic energy of the supply steam and take the exhaust steam velocity to be 120 m/s.

Calculate also the required cross-sectional area of the exhaust duct, assuming the velocity to be axial.

Answer

From tables page 8: $h_1 = 2977$ kJ/kg

page 3: $h_2 = 251 + 0.95\ (2358) = 2491.1$ kJ/kg.

Steady-flow energy equation:

$${}_1\dot{Q}_2 - {}_1\dot{W}_2 = \dot{m}\left[(h_2 - h_1) + \left(\frac{u_2^2 - u_1^2}{2}\right) + g\,(z_2 - z_1)\right]$$

$$\text{or } (-50 - 5000)\ \text{kW} = \dot{m}\left\{ (2491.1 - 2977)\ \frac{\text{kJ}}{\text{kg}} + \left(\frac{120}{2}\ \frac{\text{m}}{\text{s}}\right)^2 \left[\frac{\text{N s}^2}{\text{kg m}}\right]\left[\frac{\text{kJ}}{10^3\ \text{N m}}\right] \right\}$$

$$\text{or } \dot{m} = \frac{-5050\ \text{kJ/s}}{-478.7\ \text{kJ/kg}} = 10.55\ \text{kg/s}$$

Mass continuity: $\quad \dot{m} = \dfrac{u_2 A_2}{v_2}$

and $\quad v_2 = x_2 v_{g,2} + (1 - x_2)\, v_{f,2}$ (pp. 2 and 10)

$$= 0.95\ (7.648) + 0.05\ (0.001) = 7.2656\ \frac{\text{m}^3}{\text{kg}}.$$

(Note that $(1 - x_2)\, v_{f,2}$ is negligible!)

Thus $\quad A_2 = \dfrac{\dot{m}\, v_2}{u_2} = \dfrac{10.55\ \text{kg/s} \times 7.2656\ \text{m}^3/\text{kg}}{120\ \text{m/s}} = 0.639\ \text{m}^2.$

Example 6.5

The diagram shows the adiabatic, isobaric mixing of two streams of steam A and B. The resulting mixture is then expanded in a convergent nozzle to D.

Neglecting potential energy throughout and using the data given, calculate the temperature at D and the diameter of the nozzle at D, this being circular.

Answer

Stream A: $\dot{m}_A = 3$ kg/s; $p_A = 10$ bar;
$T_A = 200\,°C$; $u_A = 50$ m/s.
Stream B: $\dot{m}_B = 2$ kg/s; $p_B = 10$ bar;
$T_B = 250\,°C$; $u_B = 150$ m/s.
Stream C: $u_C = 100$ m/s.
Stream D: $p_D = 6$ bar; $T_D = 200\,°C$.

Mass continuity: $\dot{m}_A + \dot{m}_B = \dot{m}_C = 5\ \dfrac{\text{kg}}{\text{s}}$.

Steady-flow energy equation:

$$\dot{m}_A\,(h_A + \tfrac{1}{2}u_A^2) + \dot{m}_B\,(h_B + \tfrac{1}{2}u_B^2) = \dot{m}_C\,(h_C + \tfrac{1}{2}u_C^2) \qquad \text{(neglecting } gz \text{ terms)}$$

$$\dot{m}_A\,(h_A + \tfrac{1}{2}u_A^2) = 3\ \frac{\text{kg}}{\text{s}}\left\{2829\ \frac{\text{kJ}}{\text{kg}} + \tfrac{1}{2}\left(50\ \frac{\text{m}}{\text{s}}\right)^2\left[\frac{\text{N s}^2}{\text{kg m}}\right]\left[\frac{\text{kJ}}{10^3\ \text{N m}}\right]\right\} = 8490.75\ \text{kW}.$$

(p. 7 of tables)

Similarly, $$\dot{m}_B\,(h_B + \tfrac{1}{2}u_B^2) = 2\left[2944 + \tfrac{1}{2}\times\frac{150^2}{1000}\right] = 5910.5\ \text{kW}. \qquad \text{(p. 7)}$$

Thus $$\dot{m}_C\,(h_C + \tfrac{1}{2}u_C^2) = 8490.75 + 5910.5 = 14\,401.25\ \text{kW},$$

or $$h_C + \tfrac{1}{2}u_C^2 = \frac{14\,401.25\ \text{kW}}{5\ \text{kg/s}} = 2880.25\ \text{kJ/kg} \qquad \text{at 10 bar.}$$

Thus for $u_C = 100$ m/s;

$$h_C = 2880.25\ \text{kJ/kg} - \tfrac{1}{2}\times\frac{100^2}{1000}\ \text{kJ/kg} = 2875.25\ \text{kJ/kg},$$

or $T_C = 350\,°C$ approx. (p. 7)

For the nozzle expansion the energy equation now reads as follows:

$$h_C + \tfrac{1}{2}u_C^2 = h_D + \tfrac{1}{2}u_D^2$$

(there being no work or heat transfer in a nozzle).

Now $h_D = 2851$ kJ/kg (p. 7)

Thus $\tfrac{1}{2}u_D^2 = (h_C - h_D) + \tfrac{1}{2}u_C^2 = (2875.25 - 2851)$ kJ/kg $+ 5$ kJ/kg

$= 24.25 + 5 = 29.25$ kJ/kg

or $$u_D = \sqrt{2\times 29.25\ \frac{\text{kJ}}{\text{kg}}\left[\frac{\text{kg m}}{\text{N s}^2}\right]\left[\frac{10^3\ \text{N m}}{\text{kJ}}\right]} = 241.9\ \text{m/s}.$$

Mass continuity at D:

$$A_D = \frac{\dot{m}\,v_D}{u_D} = \frac{5\ \text{kg/s}\times 0.3532\ \text{m}^3/\text{kg}}{241.9\ \text{m/s}} = 7.3\times 10^{-3}\ \text{m}^2$$

and $$D_D = \sqrt{\frac{4}{\pi}\times A_D} = \sqrt{\frac{4}{\pi}\times 7.3\times 10^{-3}\ \text{m}^2}$$

or $D_D = 0.096$ m.

Example 6.6

A rotary compressor takes in air at low velocity from atmospheric conditions of 1 bar and 15 °C and compresses it reversibly and adiabatically to 5 bar at the rate of 5 kg/s.

If the internal diameter of the delivery pipe is 100 mm calculate the power to drive the compressor, assuming that air is a perfect gas and neglecting potential energies.

Answer

For the reversible adiabatic compression of a perfect gas,

$$T_2 = T_1\left(\frac{p_2}{p_1}\right)^{(\gamma-1)/\gamma} = 288\ \text{K}\ (5)^{0.286} = 456.3\ \text{K}.$$

Perfect gas laws at 2:

$$v_2 = \frac{RT_2}{p_2} = \frac{0.287\ \frac{\text{kJ}}{\text{kg K}} \times 456.3\ \text{K}}{500\ \frac{\text{kN}}{\text{m}^2}} = 0.2619\ \text{m}^3/\text{kg}.$$

Mass continuity at 2:

$$u_2 = \frac{\dot{m}v_2}{A_2} = \frac{5\ \frac{\text{kg}}{\text{s}} \times 0.2619\ \frac{\text{m}^3}{\text{kg}}}{\frac{\pi}{4} \times (0.1\ \text{m})^2} = 166.74\ \text{m/s}.$$

Steady-flow energy equation 1–2:

$${}_1\dot{Q}_2 - {}_1\dot{W}_2 = \dot{m}\left[(h_2 - h_1) + \frac{(u_2^2 - u_1^2)}{2} + g(z_2 - z_1)\right]$$

or ${}_1\dot{W}_2 = -\dot{m}\ [(h_2 - h_1) + \frac{1}{2}u_2^2]$ (all else is zero) $= \dot{m}\left[c_p(T_1 - T_2) - \frac{u_2^2}{2}\right]$

$$= 5\ \frac{\text{kg}}{\text{s}}\left\{1.005\ \frac{\text{kJ}}{\text{kg K}}(288 - 456.3)\ \text{K} - \frac{1}{2}\left(166.74\ \frac{\text{m}}{\text{s}}\right)^2\left[\frac{\text{N s}^2}{\text{kg m}}\right]\left[\frac{\text{kJ}}{10^3\ \text{N m}}\right]\right\}$$

$$= 5\ \frac{\text{kg}}{\text{s}}\ (-169.1 - 13.1)\ \frac{\text{kJ}}{\text{kg}} = -915.2\ \text{kW}.$$

Example 6.7

Define steady flow. Be brief.

Steam is supplied to a turbine at 100 bar, 500 °C, through a pipe 0.2 m in diameter. The exhaust steam leaves at a pressure of 0.15 bar with a dryness fraction of 0.97 through a duct 2.5 m in diameter.

The power output is 18 MW when the steam mass flow rate is 22 kg/s.

Calculate the heat transfer rate from the turbine, neglecting potential energy.

Answer

Steady flow is characterised by two things:

(a) the mass flow rate is steady,

(b) the properties across any section at right angles to the flow are invariable with time.

From tables page 8: $h_1 = 3373 \dfrac{\text{kJ}}{\text{kg}}$,

$$v_1 = 0.032\,75 \ \frac{\text{m}^3}{\text{kg}}.$$

Mass continuity $\quad u_1 = \dfrac{\dot{m} v_1}{A_1} = \dfrac{22 \ \frac{\text{kg}}{\text{s}} \times 0.032\,75 \ \frac{\text{m}^3}{\text{kg}} \times 4}{\pi \times 0.2^2} = 22.93 \ \dfrac{\text{m}}{\text{s}}.$

At exit (table p. 3), 0.15 bar is not listed and vertical interpolation is necessary; note particularly that specific volume does not vary in a linear fashion at low pressure.

$$h_{0.14,0.97} = 220 + 0.97\,(2376) = 2524.72 \text{ kJ/kg},$$

$$h_{0.16,0.97} = 232 + 0.97\,(2369) = 2529.93 \text{ kJ/kg}.$$

Thus $h_{0.15,0.97} = \dfrac{2524.72 + 2529.93}{2} = 2527.33 \text{ kJ/kg}.$

$$v_{0.14,0.97} = 0.97 \times 10.69 = 10.3693 \text{ m}^3/\text{kg},$$

(neglecting v_f)

$$v_{0.16,0.97} = 0.97 \times 9.432 = 9.149 \text{ m}^3/\text{kg}.$$

Now $(pv)_{0.14,0.97} = 0.14 \times 10.3693 = 1.4517 \text{ bar m}^3/\text{kg}$

and $(pv)_{0.16,0.97} = 0.16 \times 9.149 = 1.4638 \text{ bar m}^3/\text{kg}.$

Thus $(pv)_{0.15,0.97} = \dfrac{1.4517 + 1.4638}{2} = 1.4578 \text{ bar m}^3/\text{kg}.$

Thus $v_{0.15,0.97} = \dfrac{1.4578}{0.15} = 9.718 \text{ m}^3/\text{kg}.$ (Not the arithmetic mean of the two v's!)

Steady-flow energy equation

$${}_1\dot{Q}_2 - {}_1\dot{W}_2 = \dot{m}\left[(h_2 - h_1) + \left(\frac{u_2^2 - u_1^2}{2}\right)\right]$$

and substituting

$${}_1\dot{Q}_2 = 18 \text{ MW} + 22 \ \frac{\text{kg}}{\text{s}}\left\{(2527.33 - 3373)\ \frac{\text{kJ}}{\text{kg}} + \tfrac{1}{2}\,(43.55^2 - 22.93^2)\ \frac{\text{m}^2}{\text{s}^2}\left[\frac{\text{N s}^2}{\text{kg m}}\right]\left[\frac{\text{kJ}}{10^3 \text{ N m}}\right]\right\}\left[\frac{\text{MW}}{10^3 \text{ kW}}\right]$$

$$= 18 \text{ MW} - \frac{22}{1000}\,(-845.67 + 0.685) \text{ MW} = 18 \text{ MW} - 18.5897 \text{ MW} = 0.5897 \text{ MW}$$

(589.7 kW).

6.3 Momentum Equation in Steady Flow

Newton's law states that change of motion is proportional to impulse, and expressed mathematically

$$\delta u \propto F\,\delta t.$$

The constant of proportionality is the body mass so that for a given direction (say x),

$$(\Sigma F_x)\,\delta t = m\,\delta u_x.$$

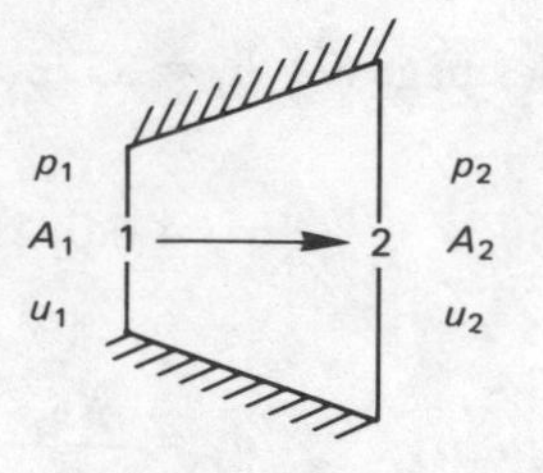

Consider a duct of varying section which is horizontal. Mass continuity shows that there will be a velocity change across the control volume and that the fluid is being accelerated, which implies a pressure change to give the necessary accelerating force.

Free-body diagram for the fluid inside the control volume:

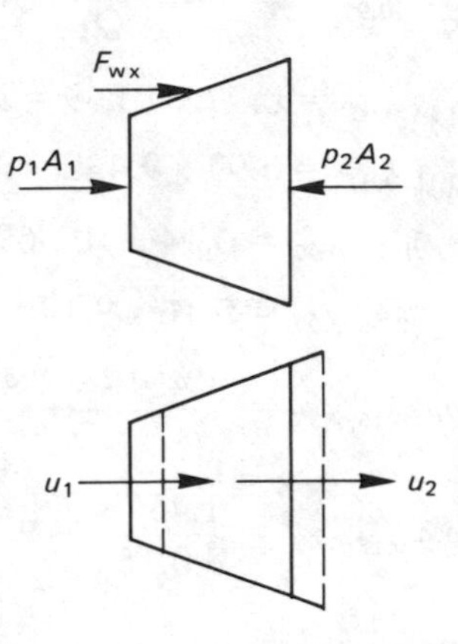

Consider the forces in the x-direction:

$$(p_1 A_1 + F_{W,x} - p_2 A_2)\,\delta t = m\,\delta u_x$$

or

$$(\Sigma F_x)\,\delta t = m\,\delta u_x$$

where F_x is the 'wall-force' component in the x-direction.

In time δt the fluid advances to a new dotted position from the solid to the dotted boundary.

For steady flow the fluid elements in the common volume are always subjected to the same forces and accelerations at any given station.

Thus the net effect is to accelerate δm from u_1 to u_2. As $\delta t \to 0$ the value of u_1 at station 1 and u_2 at station 2 are unaffected but δm reduces.

$$\Sigma F_x = \frac{\delta m}{\delta t} \times u_x = \frac{\mathrm{d}m}{\mathrm{d}t}\,\delta u_x = \dot{m}\,\delta u_x.$$

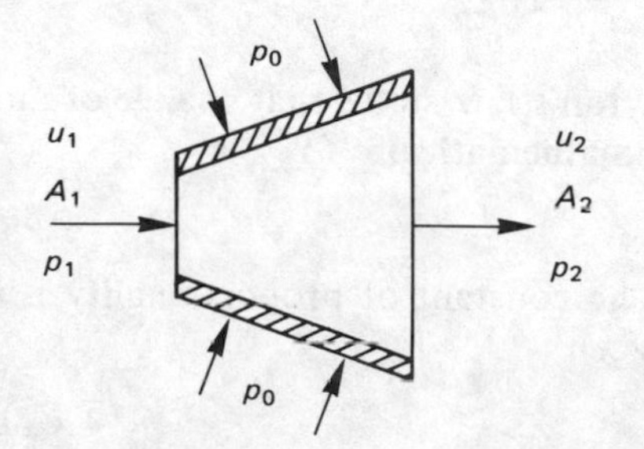

Consider now the same duct but with an external fluid at p_0.
Consider the duct walls and the fluid body inside separately pro tem.

Duct:

$$F_{\text{net},x} = F_{0,x} - F_{W,x}$$

where $F_{0,x}$ = outside fluid force in x- direction.

$$F_{0,x} = p_0 (A_2 - A_1)_x$$

where A_x = area perpendicular to flow at any given section.

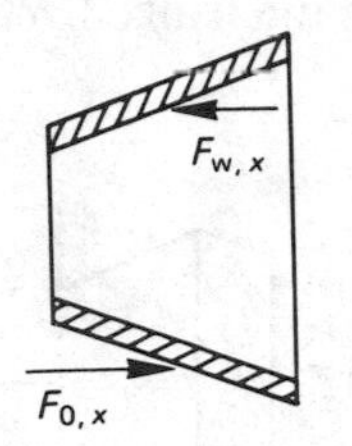

Fluid body inside duct:

Equal and opposite force $F_{W,x}$ acting on fluid

and $$\Sigma F_x = \dot{m}(u_2 - u_1)_x = p_1 A_1 + F_{W,x} - p_2 A_2 .$$

Taken together, $F_{W,x} = F_{0,x} - F_{\text{net},x} = \dot{m}(u_2 - u_1)_x + p_2 A_2 - p_1 A_1$

or, recasting, $$F_{\text{net},x} = -\dot{m}(u_2 - u_1)_x + A_1(p_1 - p_0) - A_2(p_2 - p_0).$$

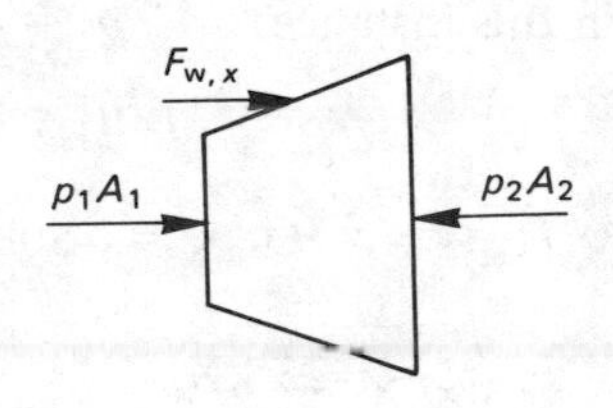

Notes

(a) The above equation can be expressed as 'The net force on the fluid in a given direction equals the rate at which momentum leaves minus the rate at which it enters in that direction.'

(b) All velocities are measured *relative to the control volume*.

(c) The equation is a *vector* equation and must be applied to all appropriate co-ordinate directions. Components of velocity and projected areas perpendicular to the chosen direction must be employed.

Thus, e.g., for flow at an angle θ to x-axis

$$\overline{u}_x = u \cos\theta$$

and force due to pressure x acceleration is

$$pA_x = pA \cos\theta \qquad \text{etc.}$$

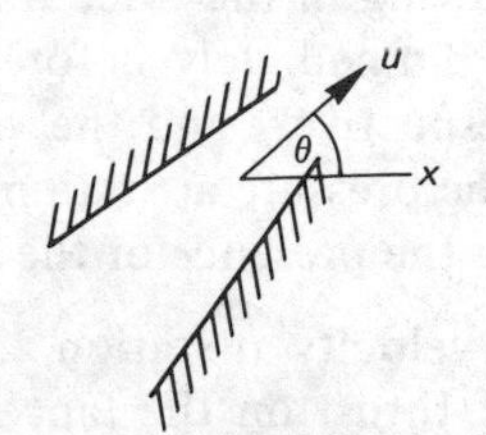

6.4 Worked Examples

Example 6.8

Air from the atmosphere is drawn into the intake of a stationary jet engine. It emerges from the propulsion nozzle at atmospheric pressure with a velocity of 500 m/s. If the mass flow through the engine is 50 kg/s, determine the engine thrust.

The engine now moves forward at 200 m/s while the velocity of the jet relative to the engine is unchanged. What is the new thrust?

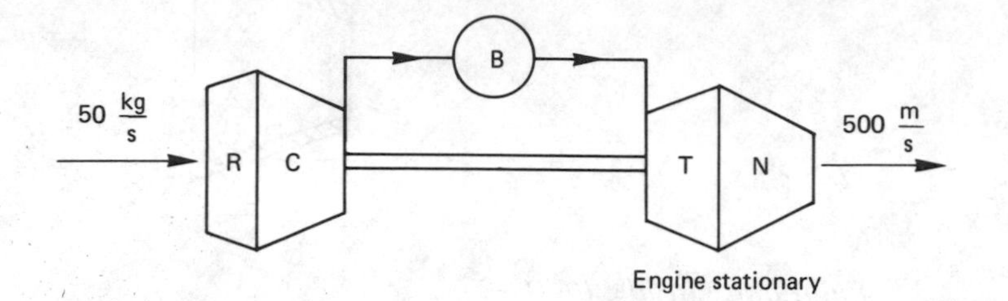

R ≡ ram intake; C ≡ compressor; B ≡ burner; T ≡ turbine; N ≡ nozzle

Answer

Generally, $F_{\text{net},x} = -\dot{m}(u_2 - u_1)_x + A_1(p_1 - p_0)_x - A_2(p_2 - p_0)_x$.

However, in this instance: $p_1 = p_2 = p_0$

and $u_1 = 0$ (assumed).

Thus $$F_{\text{net},x} = -\dot{m}u_{2,x} = -50\ \frac{\text{kg}}{\text{s}} \times 500\ \frac{\text{m}}{\text{s}} \left[\frac{\text{N s}^2}{\text{kg m}}\right] \left[\frac{\text{kN}}{10^3\ \text{N}}\right]$$

$F_{\text{net},x} = -25$ kN (right to left if positive x direction is L to R).

(This is the propulsive force trying to drive the engine forward against the braking restraint.)

For the engine moving right to left at 200 m/s with the same exhaust, again

$$p_1 = p_2 = p_0.$$

We can imagine the engine stationary once again with an equivalent u_1 of 200 m/s left to right into the ram intake.

Thus $$F_{\text{net},x} = -\dot{m}(u_2 - u_1)_x = -50\ \frac{\text{kg}}{\text{s}}\ (500 - 200)\ \frac{\text{m}}{\text{s}} \left[\frac{\text{N s}^2}{\text{kg m}}\right] \left[\frac{\text{kN}}{10^3\ \text{N}}\right]$$

$= -15$ kN (right to left).

Example 6.9

An axial-flow fan is mounted in a duct of cross-sectional area 0.2 m^2 (see diagram). The fluid flowing in the duct is air and the flow may be assumed to be adiabatic. At station 1 immediately before the fan the pressure and temperature are 1 bar and 15 °C respectively and the velocity is 30 m/s. The power input to the fan is 18 kW and the pressure at station 2 is 1.05 bar.

Neglecting the presence of the driving shaft, calculate:

(a) the fluid velocity at station 2,
(b) the axial thrust on the fan, stating its direction, if wall friction is negligible.

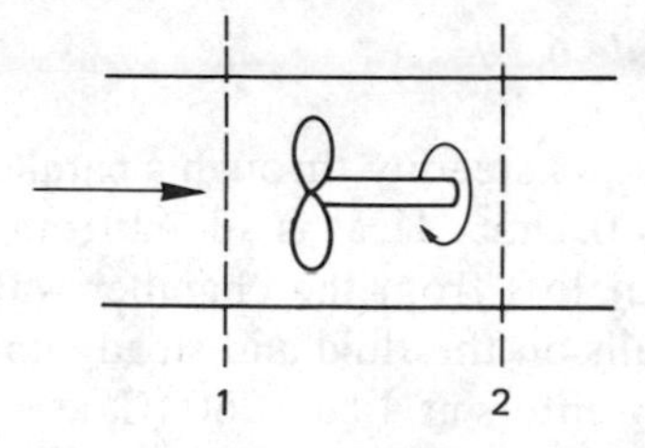

Answer

Mass continuity at 1:

$$\dot{m} = \frac{Au_1}{v_1} = \frac{Au_1 p_1}{RT_1} = \frac{0.2\ \text{m}^2 \times 30\ \dfrac{\text{m}}{\text{s}} \times 100\ \dfrac{\text{kN}}{\text{m}^2}}{0.287\ \dfrac{\text{kJ}}{\text{kg K}} \times 288\ \text{K}} = 7.259\ \frac{\text{kg}}{\text{s}}.$$

Mass continuity at 2:

$$\dot{m} = \frac{u_2 A p_2}{RT_2}$$

or
$$7.259\ \frac{\text{kg}}{\text{s}} = \frac{u_2 \times 0.2\ \text{m}^2 \times 105\ \dfrac{\text{kN}}{\text{m}^2}}{0.287\ \dfrac{\text{kJ}}{\text{kg K}} \times T_2}$$

or
$$\frac{u_2}{T_2} = 0.0992\ \frac{\text{m}}{\text{s K}} \qquad \ldots (1)$$

Steady-flow energy equation 1 → 2:

$$-{}_1\dot{W}_2 = \dot{m}\left\{c_p\,(T_2 - T_1) + \left[\frac{(u_2^2 - u_1^2)}{2}\right]\right\} \qquad \text{(all else is zero).}$$

Thus, substituting:

$$-18\ \text{kW} = 7.259\ \frac{\text{kg}}{\text{s}}\left\{1.005\ \frac{\text{kJ}}{\text{kg K}}\ (T_2 - 288)\ \text{K} + \tfrac{1}{2}(u_2^2 - 900)\ \frac{\text{m}^2}{\text{s}^2}\left[\frac{\text{N s}^2}{\text{kg m}}\right]\left[\frac{\text{kJ}}{10^3\ \text{N m}}\right]\right\}$$

or
$$-2.0297 = 1.005T_2 - 289.44 + \frac{u_2^2}{2000} \qquad \ldots (2)$$

Substituting from (1) in (2), $\quad T_2 = \dfrac{u_2}{0.0992}$,

$$-2.0297 = 1.005\ \frac{u_2}{0.0992} - 289.44 + \frac{u_2^2}{2000}.$$

This gives $u_2^2 + 20\,262.1\,u_2 - 574\,820.6 = 0$

or
$$u_2 = \frac{-20\,262.1 \pm \sqrt{(410.553 \times 10^6) + (2.2993 \times 10^6)}}{2} = 283\ \text{m/s}.$$

Momentum equation:

$$F_{\text{net},x} = -\dot{m}\,(u_2 - u_1)_x + A_{1,x}\,(p_1 - p_0) - A_{2,x}\,(p_2 - p_0)$$

$$= -\dot{m}\,(u_2 - u_1)_x + A_x\,(p_1 - p_2) \qquad (p_0 \text{ cancels and } A \text{ is constant})$$

$$= -7.259\ \frac{\text{kg}}{\text{s}}\ (28.3 - 30)\ \frac{\text{m}}{\text{s}}\left[\frac{\text{N s}^2}{\text{kg m}}\right] + 0.2\ \text{m}^2\ (100 - 105)\ \frac{\text{kN}}{\text{m}^2}\left[\frac{10^3\ \text{N}}{\text{kN}}\right]$$

$$= +\,12.34\ \text{N} - 1000\ \text{N}$$

$$= -987.7\ \text{N},$$ i.e. right to left on fan for x positive left to right.

Example 6.10

CO_2 flows steadily through a parallel-sided chamber whose constant cross-sectional area is 0.1 m². Heat is added from an external source passing entirely to the fluid without loss from the chamber walls. There are negligible drag forces exerted by the walls on the fluid and steady conditions prevail everywhere in the chamber.

CO_2 enters at 4 bar, 260 °C, at a velocity of 200 m/s and leaves at 400 m/s.

Calculate the density and pressure of the fluid leaving and the rate of heat transfer. (c_p (CO_2) = 1.102 kJ/(kg K).)

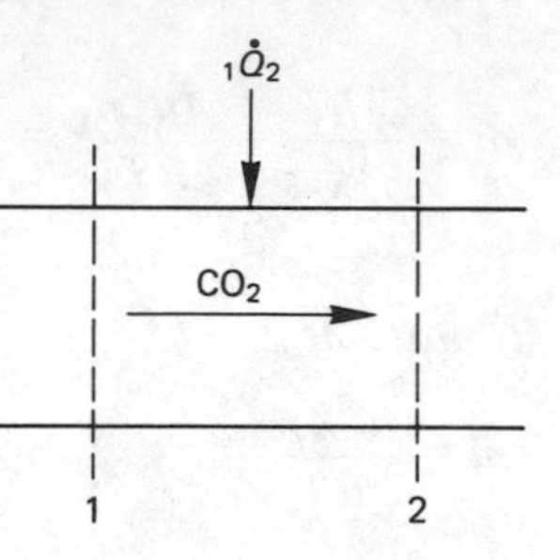

Answer

Mass continuity:

$$\rho_2 = \rho_1 \frac{u_1}{u_2} \quad (A \text{ is constant}) = \frac{p_1}{RT_1} \times \frac{u_1}{u_2} = \frac{400 \dfrac{\text{kN}}{\text{m}^2}}{0.287 \dfrac{\text{kJ}}{\text{kg K}} \times 533 \text{ K}} \times \frac{200 \dfrac{\text{m}}{\text{s}}}{400 \dfrac{\text{m}}{\text{s}}}$$

$$= 1.307 \frac{\text{kg}}{\text{m}^3}.$$

Thus $\dot{m} = \rho_2 A u_2 = 1.307 \dfrac{\text{kg}}{\text{m}^3} \times 0.1 \text{ m}^2 \times 400 \dfrac{\text{m}}{\text{s}} = 52.28 \dfrac{\text{kg}}{\text{s}}.$

Momentum equation:

$$0 = -\dot{m}(u_2 - u_1) + A(p_1 - p_2) \qquad (p_0 \text{ cancels})$$

or $$p_2 = p_1 - \frac{\dot{m}(u_2 - u_1)}{A}$$

$$= 400 \frac{\text{kN}}{\text{m}^2} - \frac{52.28 \dfrac{\text{kg}}{\text{s}} (400 - 200) \dfrac{\text{m}}{\text{s}}}{0.1 \text{ m}^2} \left[\frac{\text{N s}^2}{\text{kg m}}\right]\left[\frac{\text{kN}}{10^3 \text{ N}}\right]$$

$$= 400 - 104.56 = 295.4 \frac{\text{kN}}{\text{m}^2} \qquad (2.954 \text{ bar}).$$

$$R(CO_2) = \frac{R_0}{m_v(CO_2)} = \frac{8.3143 \dfrac{\text{kJ}}{\text{kmol K}}}{44 \dfrac{\text{kg}}{\text{kmol}}} = 0.189 \frac{\text{kJ}}{\text{kg K}}.$$

Gas laws:

$$T_2 = \frac{p_2}{R\rho_2} = \frac{295.4 \dfrac{\text{kN}}{\text{m}^2}}{0.189 \dfrac{\text{kJ}}{\text{kg K}} \times 1.307 \dfrac{\text{kg}}{\text{m}^3}} = 1196 \text{ K } (923 \text{ °C}).$$

Energy equation 1–2:

$$ {}_1\dot{Q}_2 = \dot{m}_1 q_2 = \dot{m}\left[(h_2 - h_1) + \frac{(u_2^2 - u_1^2)}{2}\right] = \dot{m}\left[c_p\,(T_2 - T_1) + \frac{(u_2^2 - u_1^2)}{2}\right] $$

$$ = 52.28\ \frac{\text{kg}}{\text{s}}\left\{1.102\ \frac{\text{kJ}}{\text{kg K}}\,(1196 - 533)\ \text{K} + \right. $$

$$ \left. + \left(\frac{400^2 - 200^2}{2}\right)\frac{\text{m}^2}{\text{s}^2}\left[\frac{\text{N s}^2}{\text{kg m}}\right]\left[\frac{\text{kJ}}{10^3\ \text{N m}}\right]\right\} $$

$$ = 52.28\,(730.6 + 60)\ \text{kW}\left[\frac{\text{MW}}{10^3\ \text{kW}}\right] = 41.33\ \text{MW}. $$

Exercises

1 Freon-12 is compressed adiabatically in steady flow from saturated vapour at 1.004 bar to 5.673 bar with 20 K degrees of superheat. The entry and exit velocities are respectively 50 m/s and 250 m/s. If the duct inlet cross-sectional area is 0.1 m^2 what is the power required? (1.949 MW)

2 Steam expands adiabatically in a nozzle from 10 bar, 250°C, to 1 bar, saturated vapour. Ignoring entry kinetic energy and potential energy everywhere calculate the exit velocity. (733.5 m/s)

3 Calculate the reactive propulsive force on the nozzle in example 6.5 (pp. 58), ignoring entry conditions and assuming that atmospheric pressure outside the nozzle exit is 1 bar. (–4.8595 kN)

4 The combustion chamber and propulsion nozzle of a rocket engine are represented in the diagram. During the test-bed firing the combustion gases expand exactly to atmospheric pressure at exit from the nozzle and leave with a velocity of 3000 m/s. Calculate the static thrust developed by the engine if the mass flow rate is 100 kg/s.

The conditions within the engine remain constant during flight, i.e. the velocity of the gases leaving the nozzle is still 3000 m/s relative to the nozzle. If the pressure of gases in the exit plane of the nozzle is 98 kN/m^2, estimate the net thrust when the rocket is travelling at an altitude of 170 km (where the atmospheric pressure is 4×10^{-3} N/m^2). The nozzle exit area is 0.6 m^2.

(300 kN, 358.8 kN)

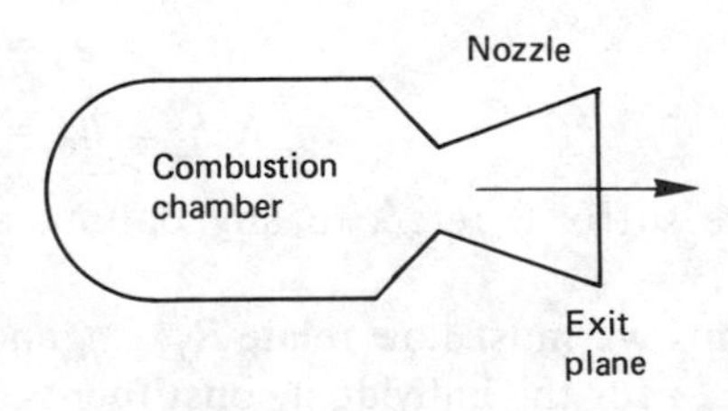

7 Mixtures

7.1 Gases

Gibbs–Dalton Laws

The treatment of mixtures is confined in this section to mixtures of perfect gases. The Gibbs–Dalton laws quoted here are absolutely correct for such mixtures and become more accurate for mixtures of real gases when the mixture pressure decreases.

The combined law states that:

> The pressure and internal energy of a mixture of gases are respectively equal to the sums of the pressures and internal energies of the individual constituents when each occupies a volume equal to that of the mixture at the mixture temperature.

In mathematical form,

$$p_t = \Sigma p_i]_{V,T}; \qquad E_t = \Sigma E_i]_{V,T},$$

where p_t = pressure of a mixture of perfect gases,
E_t = internal energy of a mixture of perfect gases,
p_i = potential pressure of any constituent gas i,
E_i = internal energy of any constituent gas i.

By extension we may also write

$$H_t = \Sigma E_i]_{V,T} + V\Sigma p_i]_{V,T}$$

whereas $$H_i = [E_i + p_i V]_{V,T} = E_i]_{V,T} + Vp_i]_{V,T},$$

i.e. $$H_t = \Sigma H_i]_{V,T}.$$

Recapitulating the laws of a perfect gas already given in Chapter 4, we may write

$$p_t v = R_t T,$$

$$e_t - e_0 = c_{v,t}(T_t - T_0)$$

$$h_t - h_0 = c_{p,t}(T_t - T_0).$$

where suffix 0 refers to any datum, and R_t, $c_{v,t}$ and $c_{p,t}$ refer to a *mixture* of gases.

Thus we must now relate R_t, $c_{v,t}$ and $c_{p,t}$ for a mixture to the values of R_i, $c_{v,i}$ and $c_{p,i}$ for the individual constituents.

Once the mixture constants can be evaluated the mixture can then be treated as if it were a single gas.

Now $$p_i V = n_i R_0 T,$$

where p_i = partial pressure of any constituent gas i,
n_i = amount of substance in volume V (in kmol).

Thus $$V\Sigma p_i = R_0 T \Sigma n_i$$

(where V and T are held constant).

But $$n_t = \Sigma n_i$$

(for any process *with no chemical change*).

Now $\quad p_t V = n_t R_0 T = m_t R_t T \qquad$ (where m_t, R_t refer to mixture).

Thus $$R_t = \frac{R_0}{m_{v,t}} = \frac{n_t}{m_t} R_0$$

Now $$n_t = \Sigma n_i = \Sigma \left(\frac{m_i}{m_{v,i}}\right)$$

and $$m_{v,i} = \frac{R_0}{R_i} = \text{relative molecular mass of constituent i.}$$

Thus $$R_t = \frac{R_0}{m_t} \Sigma m_i \frac{R_i}{R_0} = \Sigma \frac{m_i}{m_t} R_i$$

where $\dfrac{m_i}{m_t}$ = mass fraction of constituent i in total mixture.

Also $$E = \emptyset(T) \text{ only}$$

and we can dispense with suffix V for a perfect gas.

Now $$dE = mc_v \, dT$$

applies to a mixture as well as a constituent gas.

But $$E_t = \Sigma E_i]_T$$

and $$m_t c_{v,t} \, dT = \Sigma m_i \, c_{v,i} \, dT$$

where $$c_{v,t} = \Sigma \frac{m_i}{m} c_{v,i}$$

(and if respective $c_{v,i}$ values are constant $c_{v,t}$ is a constant).

Similarly,

$$c_{p,t} - c_{v,t} = R_t$$

where $c_{p,t} = \Sigma \dfrac{m_i}{m_t} R_i + \Sigma \dfrac{m_i}{m_t} c_{v,i} = \Sigma \dfrac{m_i}{m_t} c_p$.

Thus $$E_t - E_0 = \Sigma E_i]_T - \Sigma E_i]_{T_0}$$

or $$m_t(e_t - e_0) = \Sigma m_i e_i]_T - \Sigma m_i e_i]_{T_0} = \Sigma m_i c_{v,i}(T - T_0)$$

and $$m_t(h_t - h_0) = \Sigma m_i h_i]_T - \Sigma m_i h_i]_{T_0} = \Sigma m_i c_{p,i}(T - T_0)$$

(Note that entropy relations are left till Chapter 9.)

Thus for example, for a mixture of CO_2, H_2O and O_2 in gaseous form,

$$p_t = p(CO_2) + p(H_2O) + p(O_2) \qquad (p_t = \text{mixture pressure})$$

$$p_t V = n_t R_0 T$$

$$p(CO_2)V = n(CO_2)R_0 T$$

$$p(H_2O)V = n(H_2O)R_0 T$$

$$p(O_2)V = n(O_2)R_0 T$$

$$\frac{p(CO_2)}{p_t} = \frac{n(CO_2)}{n_t} = x(CO_2)$$

$x(CO_2) = \dfrac{n(CO_2)}{n_t}$ or the percentage of CO_2 by volume

$$c_{p,t} = \Sigma \frac{m_i}{m_t} c_{p,i} = \frac{\Sigma_i (nm_v c_p)}{\Sigma_i (nm_v)} = \frac{\Sigma_i (xm_v c_p)}{\Sigma_i (xm_v)}$$

(Thus we can find $c_{p,t}$ from either a mass analysis or a volume analysis.)

also
$$c_{v,t} = \frac{\Sigma_i (xm_v c_v)}{\Sigma_i (xm_v)}$$

and
$$R_t = \frac{\Sigma_i (xm_v R)}{\Sigma_i (xm_v)} = \frac{\Sigma_i (xR_0)}{\Sigma_i (xm_v)} = \frac{R_0 \Sigma_i x}{\Sigma_i (xm_v)}$$

$$R_t = \frac{R_0}{\Sigma_i (xm_v)} \qquad \text{since } \Sigma x_i = 1 \text{ by definition.}$$

7.2 Worked Examples

Example 7.1

A vessel of capacity 1 m^3 contains 5 kg of CO_2 and 2 kg of N_2. Find the mixture pressure at a vessel temperature of 20 °C.

Hence find the change in vessel pressure when the vessel temperature is raised to 50 °C.

Answer

$$R(CO_2) = \frac{R_0}{m_v(CO_2)} = \frac{8.3143 \dfrac{\text{kJ}}{\text{kmol K}}}{44 \dfrac{\text{kg}}{\text{kmol}}} = 0.189 \frac{\text{kJ}}{\text{kg K}}.$$

$$R(N_2) = \frac{R_0}{m_v(N_2)} = \frac{8.3143}{28} = 0.297 \frac{\text{kJ}}{\text{kg K}}.$$

$$p(CO_2) = \frac{m(CO_2)R(CO_2)T}{V} = \frac{5 \text{ kg} \times 0.189 \dfrac{\text{kJ}}{\text{kg K}} \times 293 \text{ K}}{1 \text{ m}^3} \left[\frac{\text{kN m}}{\text{kJ}}\right]$$

$$= 276.8 \frac{\text{kN}}{\text{m}^2} \qquad (= 2.768 \text{ bar})$$

$$p(N_2) = \frac{m(N_2)R(N_2)T}{V} = \frac{2 \times 0.297 \times 293}{1} = 174.0 \frac{\text{kN}}{\text{m}^2} \qquad (= 1.74 \text{ bar}).$$

Thus
$$p_{\text{vessel}} = 2.768 + 1.74 = 4.508 \text{ bar}.$$

For a fixed vessel volume from the gas laws:

$$p_2 = p_1 \frac{T_2}{T_1} = 4.508 \text{ bar} \times \frac{323}{293} = 4.97 \text{ bar}$$

$$\therefore \quad p_2 - p_1 = 0.462 \text{ bar}.$$

Example 7.2

The exhaust gas from an internal-combustion engine is analysed and found to contain 76 per cent N_2, 10 per cent CO_2, 14 per cent O_2 by volume. Calculate the relative molecular mass of this mixture and its density at 1.5 bar and 550 °C.

This exhaust gas is now led to a turbocharger in which it expands reversibly and adiabatically to 1 bar.

Calculate the exit gas temperature from the turbocharger. (The values of γ for N_2, CO_2, O_2 are respectively 1.4, 1.29 and 1.394.)

Answer

For the mixture,

$$R = \frac{R_0}{\Sigma x m_v} = \frac{8.3143 \dfrac{\text{kJ}}{\text{kmol K}}}{[(0.76 \times 28) + (0.1 \times 44) + (0.14 \times 32)] \dfrac{\text{kg}}{\text{kmol}}} = 0.276 \frac{\text{kJ}}{\text{kg K}}.$$

$$m_v = \frac{m}{n} = \frac{R_0}{R} = \frac{8.3143 \dfrac{\text{kJ}}{\text{kmol K}}}{0.276 \dfrac{\text{kJ}}{\text{kg K}}} = 30.12 \frac{\text{kg}}{\text{kmol}}$$

$$\rho = \frac{m}{V} = \frac{p}{RT} = \frac{150 \dfrac{\text{kN}}{\text{m}^2}}{0.276 \dfrac{\text{kJ}}{\text{kg K}} \times 823 \text{ K}} = 0.66 \frac{\text{kg}}{\text{m}^3}$$

$$\gamma_{\text{mixture}} = \frac{\Sigma_i x m_v \gamma}{\Sigma_i x m_v} = \frac{(0.76 \times 28 \times 1.4) + (0.1 \times 44 \times 1.29) + (0.14 \times 32 \times 1.394)}{(0.76 \times 28) + (0.1 \times 44) + (0.14 \times 32)}$$

$$= 1.383.$$

$$T_2 = T_2 \left(\frac{p_2}{p_1}\right)^{(\gamma-1)/\gamma} = 823 \text{ K} \left(\frac{1}{1.5}\right)^{0.383/1.383} = 735.6 \text{ K } (462.6°\text{C})$$

Example 7.3

Equal parts by volume of CO_2 and N_2 are contained in a vessel of capacity 3 m^3. Find the mass of each gas at a vessel pressure and temperature of 4 bar and 20 °C respectively.

It is now desired to change the volumetric analysis to 65 per cent CO_2 and 35 per cent N_2. How much mixture has to be removed and how much CO_2 is to be added to give this analysis if there is to be no change in the vessel pressure and temperature?

Answer

$$x(N_2) = x(CO_2) = 0.5.$$

$$p(N_2) = p(CO_2) = 0.5 p_t = 2 \text{ bar}.$$

$$R(CO_2) = \frac{R_0}{m_v} = \frac{8.3143 \dfrac{\text{kJ}}{\text{kmol K}}}{44 \dfrac{\text{kg}}{\text{kmol}}} = 0.189 \frac{\text{kJ}}{\text{kg K}}.$$

$$R(N_2) = \frac{8.3143}{28} = 0.297 \frac{\text{kJ}}{\text{kg K}}.$$

$$m(CO_2) = \left(\frac{pV}{RT}\right)(CO_2) = \frac{200 \dfrac{\text{kN}}{\text{m}^2} \times 3 \text{ m}^3}{0.189 \dfrac{\text{kJ}}{\text{kg K}} \times 293 \text{ K}} = 10.83 \text{ kg}.$$

$$m(N_2) = \left(\frac{pV}{RT}\right)(N_2) = \frac{200 \times 3}{0.297 \times 293} = 6.896 \text{ kg.}$$

Total mass in vessel = 17.726 kg.

For the new volumetric ratio (N_2 = 35 per cent),

$$m(N_2) = \left(\frac{pV}{RT}\right)(N_2) = \frac{x(N_2)\,p_t\,V}{R(N_2)\,T} = \frac{0.35 \times 400 \dfrac{\text{kN}}{\text{m}^2} \times 3 \text{ m}^3}{0.297 \dfrac{\text{kJ}}{\text{kg K}} \times 293 \text{ K}}$$

$$m(N_2) = 4.826 \text{ kg.}$$

Thus the mass of nitrogen to be removed = 6.896 – 4.826 = 2.07 kg,

and the mass of carbon dioxide removed with this = $\dfrac{10.83}{6.896} \times 2.07 = 3.25$ kg.

Thus the mass of mixture removed = 2.07 + 3.25 = 5.32 kg.

New mass

$$m(CO_2) = \left(\frac{pV}{RT}\right)(CO_2) = \left(\frac{xp_tV}{RT}\right)(CO_2) = \frac{0.65 \times 400 \times 3}{0.189 \times 293} = 14.08 \text{ kg.}$$

∴ The added mass of CO_2 = 14.08 – (10.83 – 3.25) = 6.5 kg.

Example 7.4

The reactants for a petrol engine are octane gas (C_8H_{18}) and air in the proportions 1 : 48 by volume.

The mixture is compressed and adiabatically through a volume ratio of 8.5 : 1 from initial conditions of 1 bar and 20 °C.

Calculate the final pressure and temperature and the specific work transfer during the compression. Take c_p for C_8H_{18} as 1.656, for O_2 as 0.918 and for N_2 as 1.04, all in kJ/(kg K).

Answer

The respective proportions of C_8H_{18}, O_2 and N_2 by kmol are:
C_8H_{18} = 1 kmol; O_2 = 0.21 x 48 = 10.08 kmol; N_2 = 0.79 x 48 = 37.92 kmol.
The total number of kmol = 1 + 48 = 49 kmol.

Thus

$$x(C_8H_{18}) = \frac{1}{49} = 0.0204,$$

$$x(O_2) = \frac{10.08}{49} = 0.206,$$

$$x(N_2) = \frac{37.92}{49} = 0.774.$$

$$\text{Now } R_t = \frac{R_0}{\Sigma_i x m_v} = \frac{8.3143 \dfrac{\text{kJ}}{\text{kmol K}}}{[(0.0204 \times 114) + (0.206 \times 32) + (0.774 \times 28)] \dfrac{\text{kg}}{\text{kmol}}}$$

$$= \frac{32.441}{30.590} = 0.272 \frac{\text{kJ}}{\text{kg K}};$$

also, $c_{p,t} = \dfrac{\Sigma_i(xm_v c_p)}{\Sigma_i(xm_v)}$

$$= \frac{(0.0204 \times 114 \times 1.656) + (0.206 \times 32 \times 0.918) + (0.774 \times 28 \times 1.04)}{30.59}$$

$$= 1.0605 \ \frac{\text{kJ}}{\text{kg K}};$$

and $c_{v,t} = c_{p,t} - R_t = 1.0605 - 0.272 = 0.789 \ \dfrac{\text{kJ}}{\text{kg K}}$.

Thus $\gamma_t = \left(\dfrac{c_p}{c_v}\right)_t = \dfrac{1.0605}{0.789} = 1.344;$

$$T_2 = T_1 \left(\frac{V_1}{V_2}\right)^{\gamma-1} = 293 \text{ K } (8.5)^{0.344} = 611.8 \text{ K } (338.8\ ^\circ\text{C});$$

$$p_2 = p_1 \left(\frac{V_1}{V_2}\right)^{\gamma} = 1 \text{ bar } (8.5)^{1.344} = 17.75 \text{ bar};$$

$$_1w_2 = \frac{R_t(T_1 - T_2)}{(\gamma - 1)} = \frac{0.292 \ \frac{\text{kJ}}{\text{kg K}} (293 - 611.8) \text{ K}}{0.344} = -252.07 \ \frac{\text{kJ}}{\text{kg}}$$

(i.e. work IN).

Note that in the denominator of the calculation for R_t any one of the terms has dimensions

$$\frac{\text{kmol gas}}{\text{kmol mixture}} \times \frac{\text{kg gas}}{\text{kmol gas}},$$

and total denominator gives

$$\frac{\text{kg } C_8H_{18} + \text{kg } O_2 + \text{kg } N_2}{\text{kmol mixture}} = \frac{\text{kg mixture}}{\text{kmol mixture}},$$

since masses can be added directly.

Example 7.5

A gas mixture has the following volumetric analysis:
O_2 = 25 per cent; N_2 = 50 per cent; H_2 = 25 per cent.
The mixture is at 27 °C and 2 bar.

Calculate (a) the partial pressure of each gas, (b) the density of the mixture.

In order to reduce the partial pressure of the hydrogen to one half of its initial value, air is added to the mixture. What will then be the partial pressures of the oxygen and nitrogen?

Answer

(a) $p(O_2) = (x(O_2))p_t = \dfrac{n(O_2)}{n_t} p_t = 0.25 \times 2 = 0.5$ bar;

$$p(N_2) = (x(N_2))p_t = \frac{n(N_2)}{n_t} p_t = 0.50 \times 2 = 1.0 \text{ bar};$$

$$p(H_2) = (x(H_2))p_t = \frac{n(H_2)}{n_t} p_t = 0.25 \times 2 = 0.5 \text{ bar}.$$

(b) For a mixture, $pv = RT$ where

$$R_t = \left(\frac{m_a}{m}\right) R_a + \left(\frac{m_b}{m}\right) R_b + \ldots$$

Thus, tabulating for the true gases:

Gas	n_i	$m_{v,i}$	$m_i = n_i m_{v,i}$	$\frac{m_i}{\Sigma m}$	$R_i = \frac{R_0}{m_{v,i}}$
O_2	0.25	32	8	0.3555	0.2598
N_2	0.50	28	14	0.6222	0.2969
H_2	0.25	2	0.5	0.0222	4.1571
			$\Sigma = 22.5$		

Thus

$$R_t = (0.3555 \times 0.2598) + (0.6222 \times 0.2969) + (0.0222 \times 4.1571) = 0.369 \frac{\text{kJ}}{\text{kg K}}$$

and

$$\rho = \frac{1}{v} = \frac{p}{RT} = \frac{200 \frac{\text{kN}}{\text{m}^2}\left[\frac{\text{kJ}}{\text{kN m}}\right]}{0.369 \frac{\text{kJ}}{\text{kg K}} \times 300 \text{ K}} = 1.8067 \frac{\text{kg}}{\text{m}^3}.$$

1 mol air contains 0.21 mol O_2 and 0.79 mol N_2.

To reduce partial pressure of H_2 by one half we must double the volume of the mixture.

Hence the new volumetric analysis is

		Partial pressures	
H_2	0.25		0.25 bar
N_2	(0.5 + 0.79)		1.27 bar
O_2	(0.25 + 0.21)		0.46 bar.

Example 7.6

A rigid vessel of volume 1 m^3 is connected via a pipe and valve to another rigid vessel of volume 0.1 m^3.

With the valve closed the large vessel contains air at 6 bar, 30 °C, while the smaller vessel contains nitrogen at 20 bar, 15 °C.

The valve is opened and the two fluids mix freely.

Assuming the whole system is perfectly lagged, calculate:

(a) the volumetric analysis of the mixture,
(b) the final temperature just after mixing is complete,
(c) the final pressure just after mixing is complete,
(d) the partial pressure of each constituent,
(e) the change in internal energy if the mixture is now cooled to 15°C. (c_v for air is 0.718, for O_2 is 0.658, for N_2 is 0.743, all in kJ/(kg K)).

Answer

Initial air mass in vessel:

$$m(\text{air}) = \left(\frac{pV}{RT}\right)(\text{air}) = \frac{600 \frac{\text{kN}}{\text{m}^2} \times 1 \text{ m}^3}{0.287 \frac{\text{kJ}}{\text{kg K}} \times 303 \text{ K}} = 6.9 \text{ kg}.$$

This is composed of 0.233×6.9 kg O_2 = 1.6077 kg O_2
and 0.767×6.9 kg N_2 = 5.2923 kg N_2
(using p. 24 of tables for percentage by mass).
Also, nitrogen mass in the small vessel:

$$m(N_2) = \left(\frac{pV}{RT}\right)(N_2) = \frac{2000\ \frac{\text{kN}}{\text{m}^2} \times 0.1\ \text{m}^3}{\frac{8.3143\ \frac{\text{kJ}}{\text{kmol K}}}{28\ \frac{\text{kg}}{\text{kmol}}} \times 288\ \text{K}} = 2.339\ \text{kg}.$$

Total N_2 in system = 5.2923 + 2.339 = 7.6313 kg.

Energy equation:

$$E(\text{air}) + E(N_2) = E(\text{mixture}) \qquad (Q = W = \text{etc.} = 0).$$

Thus

$$[mc_v(T - T_0)]\ (\text{air}) + [m(N_2)c_v(T - T_0)]\ (N_2) = [m(\text{air}) + m(N_2)]c_{v,t}(T_t - T_0)$$

where $c_{v,t}$, T_t refer to mixture properties
and T_0 is a datum temperature which cancels since the masses in each side of the equation are equal.

Gas	m	m_v	$\frac{m}{m_v}$	$x = (m/m_v)/\Sigma(m/m_v)$	
O_2	1.6077	32	0.0502	0.1555	(15.55%)
N_2	7.6313	28	0.2725	0.8445	(84.45%)
			Σ = 0.3227	1.0	

Thus

$$c_{v,t} = \frac{\Sigma_i(xm_vc_v)}{\Sigma_i(xm_v)}$$

$$= \frac{(0.1555 \times 32 \times 0.658) + (0.8445 \times 28 \times 0.743)}{(0.1555 \times 32) + (0.8445 \times 28)} = \frac{20.843}{28.622} = 0.728\ \frac{\text{kJ}}{\text{kg K}}.$$

Also,

$$R_t = \frac{R_0}{\Sigma_i(xm_v)} = \frac{8.3143\ \frac{\text{kJ}}{\text{kmol K}}}{28.622\ \frac{\text{kg}}{\text{kmol}}} = 0.2905\ \frac{\text{kJ}}{\text{kg K}}.$$

Thus

$$T_t = \frac{(mc_vT)(\text{air}) + (mc_vT)(N_2)}{[m(\text{air}) + m(N_2)]c_{v,t}}$$

$$= \frac{(6.9\ \text{kg} \times 0.718\ \frac{\text{kJ}}{\text{kg K}} \times 303\ \text{K}) + (2.339 \times 0.743 \times 288)\ \text{kJ}}{(6.9 + 2.339)\ \text{kg} \times 0.728\ \frac{\text{kJ}}{\text{kg K}}} = 297.6\ \text{K}\ (= 24.6\ °\text{C}).$$

Also

$$p_t = \left(\frac{mRT}{V}\right)_t = \frac{(6.9 + 2.339)\text{ kg} \times 0.2905 \dfrac{\text{kJ}}{\text{kg K}} \times 297.6\text{ K}}{1.1\text{ m}^3} = 726.1 \frac{\text{kN}}{\text{m}^2} = 7.261\text{ bar.}$$

$$\left.\begin{aligned} p(O_2) &= x(O_2)p_t = 0.1555 \times 7.261 = 1.129\text{ bar} \\ p(N_2) &= p_t - p(O_2) = 7.261 - 1.129 = 6.132\text{ bar.} \end{aligned}\right\}$$

$$\Delta E_{\text{mixture on cooling}} = m_t c_{v,t}(T_2 - T_1)$$

$$\Delta E = (6.9 + 2.339)\text{ kg} \times 0.728 \frac{\text{kJ}}{\text{kg K}} (24.6 - 15)\text{ K}$$

$$\Delta E = 64.57\text{ kJ.}$$

Exercises

1 A mixture of gases is composed of oxygen and nitrogen with a value of c_v of 0.718 kJ/(kg K). Determine the characteristic gas constant for the mixture given that the values of c_p for oxygen and nitrogen are 0.92 kJ/(kg K) and 1.04 kJ/(kg K) respectively. (0.2857 kJ/(kg K))

2 A mixture of gases has the following volumetric analysis: CH_4 = 60 per cent; CO = 20 per cent; H_2 = 10 per cent; CO_2 = 10 per cent.

Calculate the density of this mixture at 1 bar, 15 °C. If the specific heats at constant pressure of the gases are H_2 = 14.3, CH_4 = 2.2, CO = 1.05 and CO_2 = 0.834 all in kJ/(kg K) respectively, calculate the specific heats at constant pressure and constant volume for the mixture and the index of reversible adiabatic expansion. (0.906 kg/m³, 1.545 kJ/(kg K), 1.162 kJ/(kg K), 1.329)

3 At the start of expansion in an i.c. engine the products of combustion were 'frozen' and analysed and showed 12 per cent CO_2, 13 per cent H_2O and 75 per cent N_2 by volume. The expansion process was reversible and adiabatic with γ constant at the initial temperature. Given that this is 2500 K and that initial and final pressures are respectively 50 bar and 2 bar, calculate the temperature at the end of expansion. (1048.7 °C)

7.3 Gas–Vapour Mixtures

The introduction of a liquid into an evacuated vessel of greater volume than the liquid results in some evaporation if the temperature is suitable. If the latter is maintained constant an equilibrium is eventually set up between evaporation and condensation and the pressure stabilises. The maximum amount of vapour will then have been formed above the liquid for that temperature. The pressure will be the saturation vapour pressure at that temperature. If the temperature is increased more evaporation occurs and the pressure rises. Eventually, at a sufficiently high temperature all the liquid will have evaporated and the vapour will then become superheated with further rise in temperature.

It is possible to treat both gases and vapours as though they existed separately provided the total pressure is not too high. That is, Dalton's law still holds true.

7.4 Worked Examples

Example 7.7

A vessel of capacity 0.3 m^3 contains a mixture of air and steam of dryness fraction 0.6. The total pressure is 6 bar and the temperature is 130 °C.

Calçulate the mass of water, mass of saturated vapour and mass of air present.

Answer

At 130 °C $\quad v_{g,1} = 0.6686\ m^3/kg$ (p. 4 of tables)

and $\quad v_{f,1} = 0.001\ 07\ m^3/kg$. (p. 10)

Let w ≡ water; s ≡ steam; a ≡ air.

$$m_{w,1} v_{f,1} + m_{s,1} v_{g,1} = V = 0.3\ m^3. \qquad \ldots (1)$$

Also,

$$\frac{m_{s,1}}{m_{s,1} + m_{w,1}} = 0.6. \qquad \ldots (2)$$

Thus from (2) $\quad m_{s,1} = 0.6\ m_{s,1} + 0.6\ m_{w,1}$

or $\quad m_{s,1} = 1.5\ m_{w,1}$.

Thus in (1) $\quad [m_{w,1}(0.001\ 07) + m_{w,1}(1.5)(0.6686)]\ kg \times \frac{m^3}{kg} = 0.3\ m^3$

or $\quad m_{w,1} = 0.2988$ kg

and $\quad m_{s,1} = 1.5 \times 0.2988 = 0.4482$ kg.

Now $\quad p_{s,1}$ at 130 °C = 2.7 bar. (p. 4 of tables)

Thus $\quad p_{a,1} = p_{t,1} - p_{s,1} = 6$ bar $-$ 2.7 bar = 3.3 bar

and

$$m_{a,1} = \left(\frac{p_{a,1} V}{R_a T_1}\right) = \frac{330\ \frac{kN}{m^2} \times 0.3\ m^3}{0.287\ \frac{kJ}{kg\ K} \times 400\ K} = 0.8624\ kg.$$

Example 7.8

If the vessel in example 7.7 is now cooled to 100 °C calculate:

(a) the mass of vapour condensed,
(b) the final pressure in the vessel,
(c) the heat removed.

Answer

At 100 °C,

$p_{s,2} = 1.013\ 25$ bar; $v_{g,2} = 1.673\ \frac{m^3}{kg}$ (p. 2); $v_{f,2} = 0.001\ 044\ \frac{m^3}{kg}$. (p. 10 of tables)

From example 7.7, $\quad m(H_2O) = m_{s,1} + m_{w,1} = 0.2988 + 0.4482 = 0.747$ kg.

Now $\quad m_{s,2} v_{g,2} + [m(H_2O) - m_{s,2}]\ v_{f,2} = V = 0.3\ m^3$

and substituting values

$$m_{s,2}(1.673) + (0.747 - m_{s,2})(0.001\ 044) = 0.3,$$

i.e. $$m_{s,2} = \frac{0.299\,22}{1.671\,96} = 0.179 \text{ kg};$$

$$m_{w,2} = 0.747 - 0.179 = 0.568 \text{ kg};$$

and $m_{s,1} - m_{s,2}$ = mass condensed = 0.4482 − 0.179 = 0.2692 kg. . . . (1)

Now $p_{s,2}$ = 1.013 25 bar (p. 2 of tables)

and $$p_{a,2} = \frac{T_{a,2}}{T_{a,1}} p_{a,1} = \frac{373}{400} \times 3.3 \text{ bar} = 3.077 \text{ bar} \quad (V\text{constant}).$$

Thus $p_{t,2} = p_{s,2} + p_{a,2}$ = 1.013 25 + 3.077 = 4.0905 bar. . . . (2)

Energy equation:

$$_1Q_2 = m_a c_{v,a}(T_2 - T_1) + m_{s,2}e_{g,2} + m_{w,2}e_{f,2} - m_{s,1}e_{g,1} - m_{w,1}e_{f,1}$$

in the absence of work and kinetic and potential energies.

Now

$$e_{g,2} = h_{g,2} - p_{s,2}v_{g,2} = 2675.8 \text{ kJ/kg} - 101.325 \frac{\text{kN}}{\text{m}^2}\left(1.673 \frac{\text{m}^3}{\text{kg}}\right) = 2506.3 \text{ kJ/kg}.$$

(p. 2)

and

$$e_{f,2} = h_{f,2} - p_{s,2}v_{f,2} = 419.1 - 101.325\,(0.001\,044) = 419 \text{ kJ/kg}.$$

Thus

$$_1Q_2 = 0.8624 \text{ kg} \times 0.718 \frac{\text{kJ}}{\text{kg K}}(100 - 130) \text{ K} + 0.179 \text{ kg}\left(2506.3 \frac{\text{kJ}}{\text{kg}}\right)$$
$$+ 0.568 \text{ kg}\left(419 \frac{\text{kJ}}{\text{kg}}\right) - 0.4482 \text{ kg}\left(2540 \frac{\text{kJ}}{\text{kg}}\right) - 0.2988 \text{ kg}\left(546 \frac{\text{kJ}}{\text{kg}}\right)$$
$$= -18.6 + 448.6 + 238 - 1138.4 - 163.1 = -633.5 \text{ kJ}.$$

Example 7.9

A closed vessel has a volume of 3 m³ and contains air saturated with water vapour at 36 °C and a vacuum of 27 inches of mercury. The vacuum falls to 24 inches of mercury and the temperature to 24 °C.

Calculate the mass of air which has leaked into the vessel and the mass of vapour which has condensed.

The barometer reads 29.8 inches of mercury.

Answer

$$\text{Vacuum of 27 inHg} \equiv -27 \text{ inHg}\left[\frac{25.4 \text{ mmHg}}{\text{inHg}}\right]\left[\frac{\text{bar}}{750 \text{ mmHg}}\right]. \quad \text{(p. 23 of tables)}$$

Vacuum of 27 inHg = − 0.9144 bar.

$$\text{Barometric pressure} = 29.8 \times \frac{25.4}{750} = 1.009\,23 \text{ bar}.$$

$\therefore$ 27 inHg vacuum ≡ 1.0092 − 0.9144 = 0.0948 bar absolute.
$= p_{t,1}$.

*Page 4 of tables

Correspondingly, 24 inHg $\equiv \dfrac{-24 \times 25.4}{750} = -0.8128$ bar

and absolute pressure $p_{t,2} = 1.00923 - 0.8128 = 0.1964$ bar.

Now $\quad v_{g,1}$ (at $T_1 = 36\,°C) = 23.97\ \dfrac{m^3}{kg}$, (p. 2)

$$p_{s,1} = 0.0594 \text{ bar.}$$

Thus $\quad p_{a,1} = p_{t,1} - p_{s,1} = 0.0948 - 0.0594 = 0.0354$ bar.

Thus $\quad m_{s,1} = \dfrac{V}{v_{g,1}} = \dfrac{3\ m^3}{23.97\ \dfrac{m^3}{kg}} = 0.1252\ kg = m(H_2O)$

(since H_2O is *all* steam here)

and $\quad m_{a,1} = \dfrac{V}{v_{a,1}} = \dfrac{Vp_{a,1}}{R_a T_1} = \dfrac{3\ m^3 \times 3.54\ \dfrac{kN}{m^2}}{0.287\ \dfrac{kJ}{kg\ K}\ 309\ K} = 0.1198$ kg.

At $\quad T_2 = 24\,°C,\ v_{g,2} = 45.92\ \dfrac{m^3}{kg};\ p_{s,2} = 0.02982$ bar.

$$p_{a,2} = p_{t,2} - p_{s,2} = 0.1964 - 0.029\,82 = 0.166\,58 \text{ bar.}$$

$$m_{a,2} = \frac{p_{a,2}V}{R_a T_2} = \frac{16.658 \times 3}{0.287 \times 297} = 0.5862 \text{ kg.}$$

Thus $\quad m_{a,2} - m_{a,1} = 0.5862\ kg - 0.1198\ kg$

$$m_{a,2} - m_{a,1} = 0.4664 \text{ kg.}$$

Now $\quad V = m_{s,2}v_{g,2} + (m(H_2O) - m_{s,2})\,v_{f,2}$;

$$3 = m_{s,2}(45.92) + (0.1252 - m_{s,2})(0.001\,002\,76)$$

or $\quad m_{s,2} = \dfrac{2.999}{45.919} = 0.0653$ kg

and $\quad m_{s,1} - m_{s,2} = 0.1252 - 0.0653 = 0.0599$ kg.

Exercises

1 The temperature in a vessel is 36 °C and the vessel contains 0.045 kg air and 4.5 kg of saturated steam.

What is the pressure in the vessel in bar and in inches of mercury vacuum? Barometric pressure is 30 inches of mercury.

(0.0964 bar negative, 27.15 in mercury)

2 A cylinder fitted with a piston contains air saturated with water vapour. The volume is 0.3 m^3, the pressure is 3.5 bar and the temperature 60 °C. This mixture is compressed to 5.5 bar at constant temperature.

Find (a) the masses of air and H_2O present initially,
(b) the mass of H_2O condensed in the compression.

(0.01252 kg H_2O, 1.08 kg air, 0.002 kg H_2O)

8 Stoichiometry

8.1 Introduction

The general expression in kmol for combustion of a hydrocarbon fuel is

$$C_xH_y + X(O_2) + 3.76X(N_2) = a(CO_2) + b(CO) + c(H_2O) + d(H_2) + e(O_2) + f(N_2).$$

Note the following points:

(a) x and y are determined by the nature of the fuel, e.g. C_6H_6 is benzene, C_8H_{18} is octane, etc.
(b) X is the number of kmol of oxygen provided (normally as part of atmospheric air) to burn the fuel. Note that if just enough is provided to burn the carbon and hydrogen completely (and none over) the air supply is called the stoichiometric (or chemically correct) amount of air. (The term complete combustion should be used only for all combustions in which excess air is supplied, to distinguish it from the special case of stoichiometric combustion.)
(c) The ratio by volume of nitrogen to oxygen in air is 3.76 (i.e. 79 : 21 using the approximate calculations on p. 24 of tables).
(d) Generally we must allow for all the gases shown since dissociation may cause some combustion products like CO_2 to split up into O_2 and CO to some extent. However, dissociation is not considered in this because it is too advanced a topic.
(e) Nitrogen is considered *inert* (i.e. not reacting) so that in the equation above $3.76X = f$.

    Strictly, this is not so since at high temperature in practice we get oxides of nitrogen which are toxic and are the subject of stringent regulations, particularly in the United States of America. (In fact, unburnt hydrocarbons also appear to be causing trouble but all these secondary, though nevertheless important, effects are ignored in this chapter.)

The important distinction to be made in all such calculations is that while the kmol on each side of a combustion equation may not balance (e.g. $C + O_2 = CO_2$, i.e. 1 kmol C + 1 kmol O_2 gives only 1 kmol CO_2), the masses *must* balance (e.g. 12 kg C + 32 kg O_2 = 44 kg CO_2).

Thus in the original equation the kmol before combustion are

$$1(C_6H_6) + X(O_2) + 3.76X(N_2) = 1 + 4.76X \text{ total.}$$

The kmol are $a + b + c + d + e + f$ after combustion.

However, a mass balance can be written for each element as:

Carbon: $x = a + b$.

Hydrogen: $\frac{y}{2} = c + d$.

Oxygen: $X = a + \frac{b}{2} + \frac{c}{2} + e$.

Nitrogen: $3.76X = f$.

Clearly, sufficient information must be given to solve for all these variables.

Provided the temperature is high enough for the H_2O not to condense out, the products of combustion can be treated as a mixture of perfect gases and the Gibbs–Dalton laws (examined in the previous chapter) apply in the usual way.

8.2 Worked Examples

Example 8.1

Calculate the stoichiometric mass air : fuel ratio for benzene (C_6H_6).

Answer

In kmol the combustion equation may be written as

$$1(C_6H_6) + X(O_2) + 3.76X(N_2) = a(CO_2) + b(H_2O) + c(N_2).$$

(No other products in perfect combustion.)

Carbon balance: $6 = a$;
Hydrogen balance: $3 = b$;
Oxygen balance: $X = a + \frac{b}{2} = 7.5$;
Nitrogen balance: $3.76X = 28.2 = c$.

Thus
$$\frac{m_{air}}{m_{fuel}} = \frac{7.5 \text{ kmol } O_2}{\text{kmol fuel}} \left(\frac{\text{kmol fuel}}{78 \text{ kg fuel}}\right)\left(\frac{\text{kmol air}}{0.21 \text{ kmol } O_2}\right)\left(\frac{29 \text{ kg air}}{\text{kmol air}}\right)$$

$\frac{1}{m_v}$ for fuel; % by vol. air : O_2; m_v for air

or
$$\frac{m_{air}}{m_{fuel}} = 13.28 \frac{\text{kg air}}{\text{kg fuel}}.$$

In other words the above calculation is simply a dimensional exercise to get from kmol O_2/kmol fuel to kg air/kg fuel, using the appropriate terms as shown.

Example 8.2

The gravimetric composition of a liquid fuel is 0.86 kg carbon, 0.13 kg hydrogen and 0.01 kg ash. What is the mass ratio of air to liquid fuel required for stoichiometric combustion?

The fuel is burned with air and the products of combustion are sampled to determine the volumetric composition of carbon monoxide and carbon dioxide present, the results being 1 per cent CO and 10 per cent CO_2. Assuming that the hydrogen has burned completely and that the remaining undetermined products are water vapour, excess oxygen and nitrogen, calculate the actual air : fuel mass ratio. Also find the mean isobaric specific heat capacity of the products at 1200 K using data from tables page 17.

Answer

Stoichiometric combustion, in kmol per kg fuel:

$$\left(\frac{0.86}{12}\right)(C) + \left(\frac{0.13}{2}\right)(H_2) + (X)(O_2) + (3.76X)(N_2) = a(CO_2) + d(H_2O) + f(N_2)$$

(since the values of m_v are: C, 12; O_2, 32; H_2, 2; N_2, 28).

Carbon balance: $0.0717 = a.$

Hydrogen balance: $0.065 = d.$

Oxygen balance: $X = a + \frac{d}{2} = 0.1042.$

Thus
$$\frac{m_{air}}{m_{fuel}} = \frac{0.1042 \text{ kmol } O_2}{\text{kg fuel}} \left(\frac{\text{kmol air}}{0.21 \text{ kmol } O_2}\right)\left(\frac{29 \text{ kg air}}{\text{kmol air}}\right)$$

$$= 14.39 \frac{\text{kg air}}{\text{kg fuel}}.$$

Actual combustion:

$$\left(\frac{0.86}{12}\right)(C) + \left(\frac{0.13}{2}\right)(H) + (X')(O_2) + (3.76X')(N_2) = a(CO_2) + d(H_2O) + b(CO)$$
$$+ e(O_2) + f(N_2).$$

Carbon balance: $0.0717 = a + b.$

Oxygen balance: $X' = a + \frac{d}{2} + \frac{b}{2} + e.$

Hydrogen balance: $0.065 = d.$

Nitrogen balance: $3.76X' = f.$

Furthermore, from the analysis of the combustion products:

$$\frac{a+b}{100} = 0.11; \qquad \frac{a}{b} = 10.$$

Solving for the constants we get

$$11b = 0.0717$$

or
$$b = 0.006\,518$$
$$a = 0.065\,18$$
$$d = 0.065$$
$$e + f = 1 - (a + b + d) = 0.863.$$
$$4.76X' = a + \frac{b}{2} + \frac{d}{2} + e + f = 0.964 \frac{\text{kmol } O_2}{\text{kg fuel}}.$$

Now
$$f = 3.76X' = \frac{3.76}{4.76} \times 0.964 = 0.7615$$

and
$$e = 0.863 - 0.7615 = 0.1015.$$

Thus
$$\frac{m_{air}}{m_{fuel}} = \frac{0.964}{4.76} \frac{\text{kmol } O_2}{\text{kg fuel}} \left(\frac{\text{kmol air}}{0.21 \text{ kmol } O_2}\right)\left(\frac{29 \text{ kg air}}{\text{kmol air}}\right)$$
$$\frac{m_{air}}{m_{fuel}} = 27.97 \frac{\text{kg air}}{\text{kg fuel}}.$$

$$c_{p,t} = \frac{\Sigma_i n m_v c_p}{\Sigma_i n m_v}$$

$$c_{p,t} = \frac{(0.065\,18 \times 44 \times 1.28) + (0.006\,518 \times 28 \times 1.22) + (0.065 \times 18 \times 2.425) + (0.1015 \times 32 \times 1.115) + (0.7615 \times 28 \times 1.204)}{(0.065\,18 \times 44) + (0.006\,518 \times 28) + (0.065 \times 18) + (0.1015 \times 32) + (0.7615 \times 28)}$$

$$c_{p,t} = \frac{36.024}{28.790} = 1.251 \; \frac{\text{kJ}}{\text{kg K}}.$$

$$\left(\text{Note that any one term in the numerator has dimensions } \frac{\text{kmol i}}{\text{kmol t}} \times \frac{\text{kg i}}{\text{kmol i}} \times \frac{\text{kJ}}{\text{kg i K}}.\right)$$

Example 8.3

A sample of flue gas has a dry volumetric analysis of: CO_2, 10 per cent; O_2, 5 per cent; N_2, 85 per cent.

Assuming that the fuel consists entirely of carbon (C) and hydrogen (H) and that the water in the flue gas is produced entirely by combustion of the hydrogen in the fuel, estimate the proportions, by mass, of carbon and hydrogen in the fuel and also the mass air : fuel ratio.

Answer

Take 1 kg of fuel and X kg air and let A kg C + B kg H_2 make up the fuel.

Then, in kmol,

$$\left(\frac{A}{12}\right) \text{C} + \left(\frac{B}{2}\right)\text{H}_2 + \left(\frac{0.233X}{32}\right) \text{O}_2 + \left(\frac{0.767X}{28}\right) \text{N}_2 = a\text{CO}_2 + b\text{H}_2\text{O} + c\text{N}_2 + d\text{O}_2 .$$

Carbon balance: $\dfrac{A}{12} = a.$

Hydrogen (H): $\dfrac{B}{2} = b.$

Oxygen (O_2): $\dfrac{0.233X}{32} = a + \dfrac{b}{2} + d.$

Nitrogen: $\dfrac{0.767X}{28} = c.$

Also from analysis,

$$A + B = 1;$$

$$\frac{c}{a} = \frac{85}{10} = 8.5;$$

$$\frac{d}{a} = \frac{5}{10} = 0.5.$$

$$\frac{c}{a + \frac{b}{2} + d} = \frac{0.767X}{28} \times \frac{32}{0.233X} = 3.762.$$

Thus

$$\frac{8.5a}{a + \frac{B}{4} + 0.5a} = 3.762 = \frac{8.5 \frac{A}{12}}{\frac{A}{12} + \left(\frac{1-A}{4}\right) + \frac{1}{2} \times \frac{A}{12}},$$

giving

$$A = \frac{11.286}{14.143} = 0.798; B = 1 - 0.798 = 0.202.$$

Also,

$$X = \frac{28c}{0.767} = \frac{28 \times 8.5a}{0.767} = \frac{28 \times 8.5 \times 0.798}{0.767 \times 12} = 20.635 \; \frac{\text{kg air}}{\text{kg fuel}}.$$

Example 8.4

Propane (C_3H_8) is burned completely with 20 per cent excess air.

Calculate the percentage by mass of each of the constituent combustion products and the value of γ for the products as a whole treating all gases as perfect and taking a combustion temperature of 2000 K.

Answer

$$(1)\, C_3H_8 + (X)\, O_2 + (3.76X)\, N_2 = a\, CO_2 + b\, H_2O + c\, N_2$$

for stoichiometric combustion.

Carbon balance: $3 = a.$

Hydrogen balance: $8 = 2b.$

Oxygen balance: $X = a + \frac{b}{2} = 5.$

Nitrogen balance: $3.76X = c = 18.8.$

Thus for actual combustion $X' = 1.2X = 6$ and the equation reads

$$(1)\, C_3H_8 + (6)O_2 + (22.56)N_2 = (3)CO_2 + (4)H_2O + (1)O_2 + (22.56)N_2.$$

Gas	n_i	$m_{v,i}$	$n_i m_{v,i}$	$\frac{n_i m_{v,i}}{\Sigma_i n m_v}$
CO_2	3	44	132	0.152 = 15.2% by mass
H_2O	4	18	72	0.083 = 8.3% by mass
O_2	1	32	32	0.037 = 3.7% by mass
N_2	22.56	28	631.68	0.728 = 7.28% by mass
			$\Sigma = 867.68$	

At 2000 K,

$$c_p(CO_2) = 1.371\ \frac{\text{kJ}}{\text{kg K}};\ c_p(H_2O) = 2.836\ \frac{\text{kJ}}{\text{kg K}};$$

$$c_p(O_2) = 1.181\ \frac{\text{kJ}}{\text{kg K}};\ c_p(N_2) = 1.284\ \frac{\text{kJ}}{\text{kg K}};$$

(from p. 17 of tables)

Thus

$$c_{p,t} = \Sigma_i m c_p = \left(0.152\ \frac{\text{kg } CO_2}{\text{kg products}} \times 1.371\ \frac{\text{kJ}}{\text{kg } CO_2\ \text{K}}\right)$$
$$+ \left(0.083\ \frac{\text{kg } H_2O}{\text{kg products}} \times 2.836\ \frac{\text{kJ}}{\text{kg } H_2O\ \text{K}}\right)$$
$$+ \left(0.037\ \frac{\text{kg } O_2}{\text{kg products}} \times 1.181\ \frac{\text{kJ}}{\text{kg } O_2\ \text{K}}\right)$$
$$+ \left(0.728\ \frac{\text{kg } N_2}{\text{kg products}} \times 1.284\ \frac{\text{kJ}}{\text{kg } N_2\ \text{K}}\right)$$
$$= 1.422\ \frac{\text{kJ}}{\text{kg products K}}.$$

Similarly, $R_t = \frac{\Sigma_i mR}{\Sigma_i m} = \frac{\Sigma_i n m_v R}{\Sigma_i n m_v} = \frac{R_0 \Sigma_i n}{\Sigma_i n m_v}$

$$= \frac{8.3143\ \frac{\text{kJ}}{\text{kmol K}}\ (3 + 4 + 1 + 22.56)\ \text{kmol}}{867.68\ \text{kg products}} = 0.293\ \frac{\text{kJ}}{\text{kg products K}}.$$

Thus $\quad c_{v,t} = c_{p,t} - R_t = 1.422 - 0.293 = 1.129 \dfrac{\text{kJ}}{\text{kg products K}}$

and $\quad \gamma_t = \dfrac{c_{p,t}}{c_{v,t}} = \dfrac{1.422}{1.129} = 1.259.$

Example 8.5

Ethanol (C_2H_5OH) is burned with 10 per cent excess air. Determine the mass air : fuel ratio and the volumetric analysis of the dry products of combustion assuming this is complete.

What gases would you expect to find in the products of combustion if there was insufficient air for combustion?

Answer

Let X kmol of air be supplied per kmol ethanol for stoichiometric combustion. The equation for actual combustion then reads:

$$(1)C_2H_5OH + (1.1 \times 0.21X)O_2 + (1.1 \times 0.79X)N_2 = aCO_2 + bH_2O + dO_2 + eN_2.$$

Carbon balance: $\quad 2 = a.$

Hydrogen balance: $\quad 6 = 2b.$

Oxygen balance: $\quad (1.1 \times 0.21X) + 0.5 = a + \dfrac{b}{2} + d.$

Nitrogen balance: $\quad 1.1 \times 0.79X = e.$

For stoichiometric combustion $d = 0$ and 1.1 becomes 1.0.

Then
$$0.21X = 2 + \frac{3}{2} - \frac{1}{2} = 3$$

or
$$X = 14.2857 \frac{\text{kmol air}}{\text{kmol fuel}}.$$

$$\therefore \textit{Actual} \; \frac{m_{\text{air}}}{m_{\text{fuel}}} = 1.1 \times 14.2857\left(\frac{\text{kmol air}}{\text{kmol fuel}}\right) \times \left(\frac{29 \text{ kg air}}{\text{kmol air}}\right)\left(\frac{\text{kmol fuel}}{46 \text{ kg fuel}}\right)$$

$$\frac{m_{\text{air}}}{m_{\text{fuel}}} = 9.91 \frac{\text{kg air}}{\text{kg fuel}}.$$

$$d = \left[1.1 \times 0.21\left(\frac{3}{0.21}\right)\right] + 0.5 - 3.5 = 0.3 \text{ kmol},$$

$$e = \left[1.1 \times 0.79 \times \left(\frac{3.}{0.21}\right)\right] = 12.414 \text{ kmol}.$$

Dry analysis:

Gas	n_i	$\dfrac{n_i}{\Sigma_i n}$	
CO_2	2.0	0.1359 =	13.59%
O_2	0.3	0.0204 =	2.04%
N_2	12.414	0.8437 =	84.37%
	$\Sigma n_i = 14.714$	1.00	100.00

With insufficient air for combustion not all C will burn to CO_2 and not all H_2 will burn to H_2O. Thus combustion gases will contain CO_2, CO, H_2O, H_2 and N_2 (neglecting dissociation).

Example 8.6

The chemical analysis of a petrol sample was 86 per cent carbon and 14 per cent hydrogen by mass.

After the sample was burnt with air in an internal-combustion engine the analysis of the dry products of combustion showed 15 per cent by volume of carbon dioxide and some oxygen.

Calculate the mass air : fuel ratio actually supplied to the engine and the mixture strength.

Answer

Let X kg oxygen be supplied per kg fuel. Then

$$C + O_2 = CO_2 \qquad 2H_2 + O_2 = 2H_2O$$

$$12\text{ kg} + 32\text{ kg} = 44\text{ kg}; \qquad 4\text{ kg} + 32\text{ kg} = 36\text{ kg}.$$

$$\text{Stoichiometric oxygen needed} = \underset{\text{carbon}}{\left(0.86 \times \frac{32}{12}\right)} + \underset{\text{hydrogen}}{\left(0.14 \times \frac{32}{4}\right)} = \frac{3.413\text{ kg}}{\text{kg fuel}}.$$

Excess oxygen in dry exhaust gas = $X - 3.413$ kg per kg fuel.

$$\text{Nitrogen in dry exhaust gas} = X \times \frac{0.7553}{0.2314} = 3.264X\text{ kg per kg fuel.}$$

$$\text{Carbon dioxide in dry exhaust gas} = 0.86 \times \frac{44}{12} = 3.153\text{ kg per kg fuel.}$$

Tabulate:

Gas	m_i	$m_{v,i}$	$n_i = \frac{m_i}{m_{v,i}}$
CO_2	3.153	44	0.071 7
O_2	X-3.143	32	$0.031\,25X - 0.106\,7$
N_2	$3.264X$	28	$0.116\,57X$
			$\Sigma n_i = 0.147\,82X - 0.035$

$$\text{Now } x(CO_2) = \frac{n(CO_2)}{\Sigma n} = 0.15 = \frac{0.0717}{0.147\,82X - 0.035}$$

or $X = 3.47$ kg O_2 per kg fuel

$$\text{and } \frac{m_{air}}{m_{fuel}} = \frac{3.47}{0.2314} = 14.99\,\frac{\text{kg air}}{\text{kg fuel}}.$$

$$\text{Stoichiometric } \frac{m_{air}}{m_{fuel}} = \frac{0.413}{0.2314} = 14.749\,\frac{\text{kg air}}{\text{kg fuel}}.$$

$$\text{Thus mixture strength} = \frac{14.749}{14.99} = 0.984.$$

Example 8.7

Octane vapour (C_8H_{18}) is burned with air in the mass ratio of 1 : 12.5.

Calculate the volumetric analysis of the wet products of combustion and the mixture strength.

Let X kmol O_2 be supplied per kmol C_8H_{18}.

Now $\quad C + O_2 = CO_2$; $\quad 2H_2 + O_2 = 2H_2O$.

Actual combustion equation in kmol is

$$1(C_8H_{18}) + X(O_2) + 3.76X(N_2) = a(CO_2) + b(CO) + c(H_2O) + d(N_2).$$

Carbon balance: $\quad 8 = a + c.$

Hydrogen balance: $\quad 9 = c.$

Oxygen balance: $\quad X = a + \frac{b}{2} + \frac{c}{2}.$

Nitrogen balance: $\quad 3.76X = d.$

Now $\quad X \frac{\text{kmol } O_2}{\text{kmol fuel}} \times \frac{\text{kmol air}}{0.2095 \text{ kmol } O_2} \times \frac{29 \text{ kg air}}{\text{kmol air}} \times \frac{\text{kmol fuel}}{114 \text{ kg fuel}} = 12.5 \frac{\text{kg air}}{\text{kg fuel}}.$

Thus $X = 10.294 \frac{\text{kmol } O_2}{\text{kmol fuel}}$.

Thus $\quad d = 3.76 \times 10.294 = 38.707 \text{ kmol},$

$c = 9 \text{ kmol},$

$11.371 = a + 4.5 + \frac{c}{2}.$

Also $\quad 8 = a + b$

and $\quad 10.294 = a + \frac{b}{2} + \frac{c}{2}$, giving $2.294 = \frac{c - b}{2} = 4.5 - \frac{b}{2}$

or $\quad b = 4.412 \text{ kmol}$

and $\quad a = 8 - 4.412 = 3.588 \text{ kmol}.$

Thus $\quad \Sigma$ wet exhaust gas $= a + b + c + d = 55.705$ kmol

and

$$x(CO_2) = \frac{a \times 100}{\Sigma \text{ w.e.g.}} = \frac{358.8}{55.705} = 6.44 \text{ per cent},$$

$$x(CO) = \frac{b \times 100}{\Sigma \text{ w.e.g.}} = \frac{441.2}{55.705} = 7.92 \text{ per cent},$$

$$x(H_2O) = \frac{c \times 100}{\Sigma \text{ w.e.g.}} = \frac{900}{55.705} = 16.16 \text{ per cent},$$

$$x(N_2) = 100 - (6.44 + 7.92 + 16.16) = 69.48 \text{ per cent}.$$

The stoichiometric oxygen supply is obtained from the stoichiometric equation

$$C_8H_{18} + 12.5(O_2) + (12.5 \times 3.76)(N_2) = 8(CO_2) + 9(H_2O) + (12.5 \times 3.76)(N_2)$$

or $\quad$ stoichiometric O_2 supply $= 12.5 \frac{\text{kmol } O_2}{\text{kmol fuel}}$ $\quad$ and

stoichiometric $\frac{m_{air}}{m_{fuel}} = 12.5 \left(\frac{\text{kmol } O_2}{\text{kmol fuel}}\right) \left(\frac{29 \text{ kg air}}{\text{kmol air}}\right) \left(\frac{\text{kmol fuel}}{114 \text{ kg fuel}}\right) \left(\frac{\text{kmol air}}{0.2095 \text{ kmol}}\right.$

i.e. $\quad$ stoichiometric $\frac{m_{air}}{m_{fuel}} = 15.18 \frac{\text{kg air}}{\text{kg fuel}}.$

Thus $\quad$ mixture strength $= \frac{\text{stoichiometric mass air : fuel ratio}}{\text{actual mass : fuel ratio}}$

$$= \frac{15.18}{12.5} = 1.214.$$

Exercises

1 Propane (C_3H_8) is burned with 60 per cent excess air by volume. Determine the gravimetric and volumetric air : fuel ratios.

What would be the volumetric analysis of the dry products of combustion, assuming the combustion is complete?

(Mass ratio 25.17 volumetric ratio 38.2; CO_2, 8.3 per cent; O_2, 8.3 per cent; N_2, 83.4 per cent)

2 A partial analysis of the dry exhaust gas by volume in a gas-fired boiler indicates CO_2, 9 per cent; CO, 1 per cent.

The remainder may be assumed to be nitrogen and excess oxygen. The fuel is natural gas, which may be taken as 100 per cent methane (CH_4). Calculate the volumetric and gravimetric air : fuel ratios used in the combustion process.

(Volumetric ratio 10.95 : 1, gravimetric ratio 19.85 : 1)

3 In an engine test the dry exhaust gas analysis was CO_2, 12 per cent; O_2, 3.2 per cent; N_2, 84.8 per cent. The fuel was a pure hydrocarbon. Calculate the ratio of C to H in the fuel, the mass air : fuel ratio used and the mixture strength defined as m.s. = (stoichiometric mass A/F ratio)/(Actual mass A/F ratio).

(C, 83.1 per cent; H_2, 16.9 per cent; 17.866, 85.7 per cent)

9 Second Law of Thermodynamics—Entropy

9.1 Introduction

We have seen in Chapter 3 how work transfer in a *reversible* process may be expressed as:

$$\delta W_{rev} = p\,dV. \qquad \ldots(1)$$

The second law provides a corresponding expression for heat transfer:

$$\delta Q_{rev} = T\,dS. \qquad \ldots(2)$$

where S is the *entropy* of the fluid.

Entropy follows from the second law in much the same way as internal energy follows from the first law. Unlike p and V, which are experimentally measurable, entropy is an abstract but nevertheless vital property of the fluid. If there is difficulty in understanding the property it probably lies in the fact that it can only be calculated but not measured.

Pressure gauges and scales are readily understood but an abstract quality like entropy is 'elusive' and less 'attractive' because of its abstract nature. It is therefore very important to show its relevance right at the start.

Without entropy no real engineering calculations in this subject can be performed. Any real process in engineering involves a net increase in entropy between the system and its surroundings and the size of this increase is to some scale a measure of the irreversibility of the process – i.e. its departure from the ideal reversible case. In other words, the larger the entropy gain in the process the more expensive the process becomes.

A considerable amount of lecture material is necessary to introduce entropy to a student covering heat engines, thermodynamic temperature, reversibility, and so on, and when one finally arrives at an abstract quantity – entropy – the student is prone to wonder where it is all leading.

One vital example may perhaps show its importance.

In equation (2) if $\delta Q_{rev} = 0$ we have by definition a reversible adiabatic process.

Also $T\,dS = 0$ and since $T \neq 0$, $dS = 0$, i.e. S is constant; i.e. a reversible adiabatic is an *isentrope*.

Consider the case of both ideal reversible adiabatic and actual irreversible adiabatic expansion in a turbine.

The temperature–entropy diagram overleaf shows the following:

(a) 1–2 Reversible adiabatic i.e. isentropic expansion from p_1 to p_2.
(b) 1–2_S Real, irreversible, adiabatic expansion from p_1 to p_2.

No calculation to find point 2 can be attempted in practice without a knowledge of entropy.

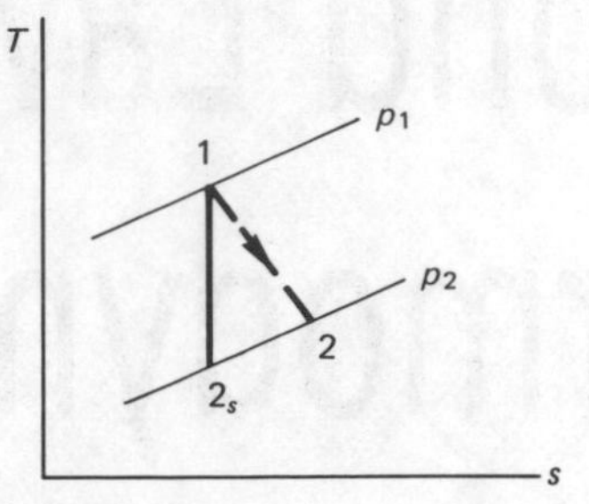

The general procedure is to find 2_s using $s_1 = s_{2,s}$ and then to calculate ${}_1w_{2,s}$ from the steady-flow energy equation.

Then, by referring to already established practice, an *efficiency* of expansion is selected which we call isentropic efficiency:

$$\eta = \frac{h_1 - h_2}{h_1 - h_{2,s}}$$

so that $h_{2,s}$ can be found (and thus $T_{2,s}$ etc).

(Or, vice versa, we measure ${}_1w_2 = h_1 - h_2$ and compare it with ${}_1w_{2,s}$ and then calculate η.)

The point to be stressed is that without entropy no calculation is possible and since all real processes are subject to irreversibility entropy must be mastered.

Certain consequential results follow from the equation (2) and previously derived theory.

Thus per unit mass

$$T\,\mathrm{d}s = \delta q_{\mathrm{rev}} = \mathrm{d}e + p\,\mathrm{d}v = \mathrm{d}h - v\,\mathrm{d}p. \qquad \ldots (3)$$

(3) is virtually an 'illuminated address' for this subject for it combines both first and second laws together to provide certain very valuable results.

Thus $$\mathrm{d}s = \frac{\mathrm{d}e}{T} + \frac{p\,\mathrm{d}v}{T} = \frac{\mathrm{d}h}{T} - \frac{v\,\mathrm{d}p}{T}.$$

In the case of a perfect gas

$$\mathrm{d}e = c_v\,\mathrm{d}T; \qquad \mathrm{d}h = c_p\,\mathrm{d}T; \qquad pv = RT.$$

Thus $$\mathrm{d}s = \frac{c_v\,\mathrm{d}T}{T} + \frac{R\,\mathrm{d}v}{v}; \quad \text{and} \quad \mathrm{d}s = \frac{c_p\,\mathrm{d}T}{T} - \frac{R\,\mathrm{d}p}{p}$$

and $$s_2 - s_1 = c_v \ln\frac{T_2}{T_1} + R\ln\frac{v_2}{v_1} = c_p \ln\frac{T_2}{T_1} - R\ln\frac{p_2}{p_1}. \qquad \ldots (4)$$

Reversible adiabatic expansion of a gas (or compression):

Here $$\delta Q_{\mathrm{rev}} = T\,\mathrm{d}S = \mathrm{d}E + p\,\mathrm{d}V = 0$$

or, per kg, $$\frac{\mathrm{d}e}{T} + \frac{p\,\mathrm{d}v}{T} = 0,$$

i.e. $$c_v\frac{\mathrm{d}T}{T} + R\frac{\mathrm{d}v}{v} = 0$$

and integrating, $$c_v \ln\frac{T_2}{T_1} + R\ln\frac{v_2}{v_1} = 0.$$

Thus $$\ln\frac{T_2}{T_1} = -\frac{R}{c_v}\ln\frac{v_2}{v_1} = \frac{R}{c_v}\ln\frac{v_1}{v_2} = (\gamma - 1)\ln\frac{v_1}{v_2}$$

(since $R = c_p - c_v$).

Thus $\quad \dfrac{T_2}{T_1} = \left(\dfrac{v_1}{v_2}\right)^{\gamma-1} \quad$ and using $\dfrac{p_1 v_1}{T_1} = \dfrac{p_2 v_2}{T_2}$

we get $\quad \dfrac{T_2}{T_1} = \left(\dfrac{p_2}{p_1}\right)^{(\gamma-1)/\gamma} \quad \ldots (5)$

and $\quad p_1 v_1^\gamma = p_2 v_2^\gamma$.

(Note that this result can be argued directly from first law $\delta Q_{rev} = dE + p\,dV = 0$ without reference to S at all.)

One other factor of importance needs stressing, which is that just as the area under a curve on the *p–v* field represents *work*, so the area under a curve on the *T–s* field represents *heat* for a reversible process (see diagram).

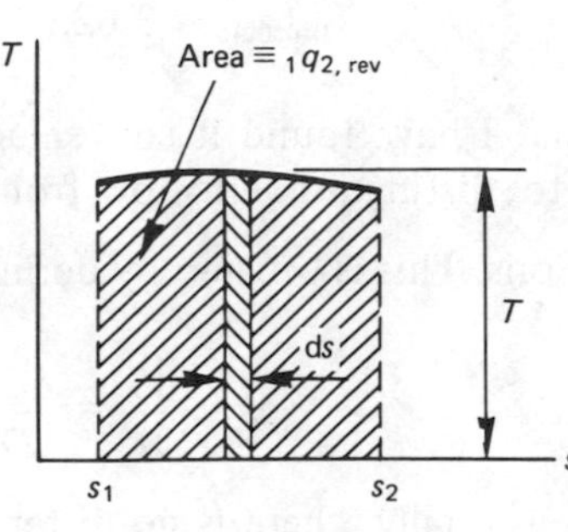

Typical *T–s* and *h–s* diagrams for a pure substance are given here.

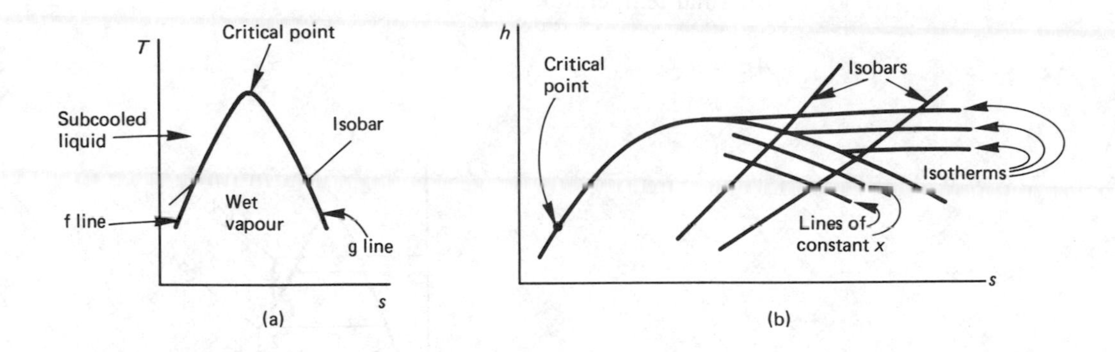

On the *T–s* field the liquid (f) and vapour (g) lines are symmetrical about a vertical line and merge at the critical point. A typical isobar is shown.

On the *h–s* field the boundary between liquid and vapour states is displaced as shown and liquid states are not generally shown on the abbreviated versions on sale.

Note that since the slope of an isobar is given by dh/ds and from previous theory $T\,ds = dh - v\,dp$. Thus $(dh/ds)_p = T$ for an isobar.

Thus when the temperature T is constant as in evaporation the isobar will be a straight line whose slope increases with the value of T and will change to a curve (concave upwards) in the superheat region as T rises again here.

The first step in familiarising yourself with entropy is to learn how to use tables of properties to find the value of s for a vapour.

9.2 Worked Examples

Example 9.1

Calculate the specific entropy for H_2O at 140 bar, 0.46 dry.

Answer

Note that s follows the same rules for dryness as do specific internal energy, specific enthalpy and specific volume, i.e.

$$s = s_f + x s_{fg}.$$

Thus on page 5 of tables,

$$s_{140, 0.46} = 3.623 + 0.46\,(1.750) = 4.428\ \frac{\text{kJ}}{\text{K kg}}.$$

Note that I have found it convenient to write (K kg) in that order in the denominator to distinguish entropy from specific heat capacity which has the same dimensions. This is because by definition $\mathrm{d}S = \dfrac{\delta Q_{rev}}{T}$ $\left(\text{in } \dfrac{\text{kJ}}{\text{K}}\right)$

and $$\mathrm{d}s = \frac{\mathrm{d}S}{m} = \frac{\delta Q_{rev}}{Tm} \left(\frac{\text{kJ}}{\text{K kg}}\right)$$ in that order.

Dimensionally, there is no difference between c_p, c_v and s but the distinction (only used in this book!) may be of help to you.

Example 9.2

One kg steam at 200 bar, 375 °C, is expanded isentropically to 100 bar. What is the final temperature?

Answer

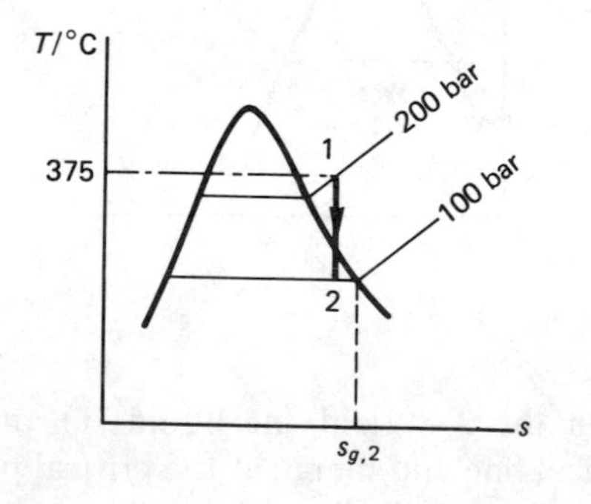

$$s_1 = s_2 = 5.228\ \frac{\text{kJ}}{\text{K kg}}. \quad \text{(p. 8 of tables)}$$

Now at 100 bar $s_{g,2} = 6.615\ \dfrac{\text{kJ}}{\text{K kg}}$.

Thus H_2O is wet vapour at 2 and $T = T_{sat} = 311$ °C.

Note that $x_2 = \dfrac{s_2 - s_{f,2}}{s_{fg,2}} = \dfrac{5.228 - 3.360}{2.255} = 0.828.$

Example 9.3

1 kg of steam is compressed adiabatically and irreversibly from 2 bar, saturated vapour to 4 bar, 200 °C. Calculate the change in specific entropy.

Answer

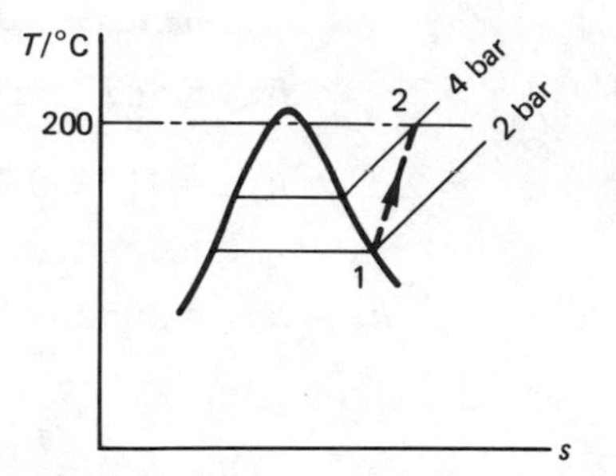

$$s_1 = 7.127 \frac{\text{kJ}}{\text{K kg}}. \quad \text{(p. 4 of tables)}$$

$$s_2 = 7.172 \frac{\text{kJ}}{\text{K kg}}. \quad \text{(p. 6)}$$

Thus $s_2 - s_1 = 0.045 \dfrac{\text{kJ}}{\text{K kg}}$.

Note that we can only show this process as a dotted line on the temperature–entropy field since in an irreversible process the intermediate states are unknown.

Example 9.4

1 kg steam expands adiabatically and irreversibly through a nozzle in steady flow from 10 bar, 300 °C, to 1 bar with an isentropic efficiency of 0.9. Calculate the exit velocity, ignoring the ongoing kinetic energy.

Answer

In steady flow,

$${}_1\dot{Q}_2 = {}_1\dot{W}_2 = \Delta gz = \frac{u_1^2}{2} = 0 \qquad \text{(for a nozzle as given).}$$

Thus
$$\frac{u_2^2}{2} = h_1 - h_2$$

or
$$u_2 = \sqrt{2(h_1 - h_2)}.$$

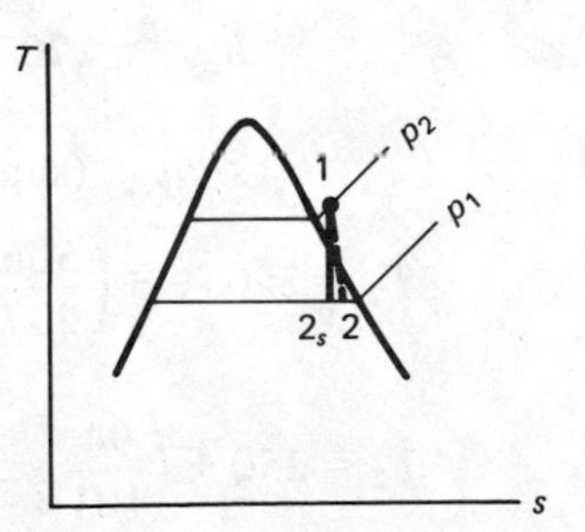

Now $$h_1 = 3052 \frac{\text{kJ}}{\text{kg}}, \quad \text{(p. 7 of tables)}$$

$$s_1 = 7.124 \frac{\text{kJ}}{\text{K kg}} = s_{2,s}$$

$$x_{2,s} = \frac{s_2 - s_{f,2}}{s_{fg,2}} = \frac{7.124 - 1.303}{6.056} = 0.961. \qquad \text{(p. 4)}$$

Thus $$h_{2,s} = h_{f,2} + x_{2,s}(h_{fg,2})$$

$$h_{2,s} = 417 + 0.961\,(2258) = 2587.4 \frac{\text{kJ}}{\text{kg}}$$

and $$h_1 - h_{2,s} = 3052 - 2587.4 = 464.6 \frac{\text{kJ}}{\text{kg}}.$$

Now, $$\eta_{\text{nozzle}} = \frac{h_1 - h_2}{h_1 - h_{2,s}} \qquad \left(\text{i.e. } \frac{u_2^2}{u_{2,s}^2}\right)$$

(since η_{nozzle} compares the k.e. actually generated with that ideally generated, i.e. when $\eta = 1.0$).

Thus $$h_1 - h_2 = 0.9\,(464.6) = 418.14 \frac{\text{kJ}}{\text{kg}}$$

and $$u_2 = \sqrt{2\,(418.14) \frac{\text{kJ}}{\text{kg}} \left[\frac{\text{kg m}}{\text{N s}^2}\right] \left[\frac{\text{kN m}}{\text{kJ}}\right]} = 914.5 \frac{\text{m}}{\text{s}}.$$

Example 9.5

1 kg of steam is throttled from 5 bar, 300 °C, to 2 bar.

Calculate the temperature after throttling and the change in specific entropy.

Answer

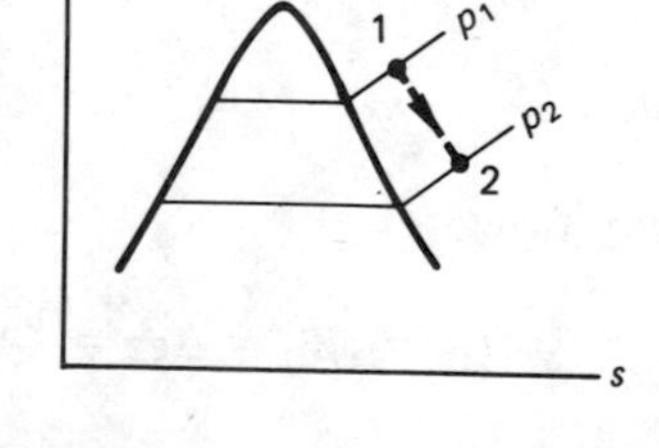

In throttling, $$h_1 = h_2 \quad (\text{since } {}_1\dot{Q}_2 = {}_1\dot{W}_2 = \Delta u^2 = \Delta gz = 0)$$

and here $$h_1 = h_2 = 3065 \frac{\text{kJ}}{\text{kg K}}. \quad \text{(p. 7 of tables)}$$

Now $$h_{g,2} = 2707 \frac{\text{kJ}}{\text{kg}}, \quad \text{(p. 4)}$$

$$\therefore \quad h_2 > h_{g,2} \quad \text{(superheated at 2)}$$

and $$T_2 \doteq 250\,°\text{C} + \left(\frac{3065 - 2971}{3072 - 2971}\right)(300 - 250)\text{ K} \quad \text{(p. 6)}$$

$$T_2 = 250 + \left(\frac{94}{101} \times 50\right) = 296.5\,°\text{C}.$$

$$s_1 = 7.460 \frac{\text{kJ}}{\text{K kg}}; \quad s_2 = 7.708 + \frac{94}{101}(7.892 - 7.708) = 7.8792 \frac{\text{kJ}}{\text{K kg}}.$$

Thus $$s_2 - s_1 = 7.879 - 7.460 = 0.419 \frac{\text{kJ}}{\text{K kg}}.$$

Example 9.6

1 kg steam at 2.7 bar, 0.5 dry, undergoes a reversible isothermal heat addition of 1282.3 kJ. What is the change of entropy and the final pressure?

Answer

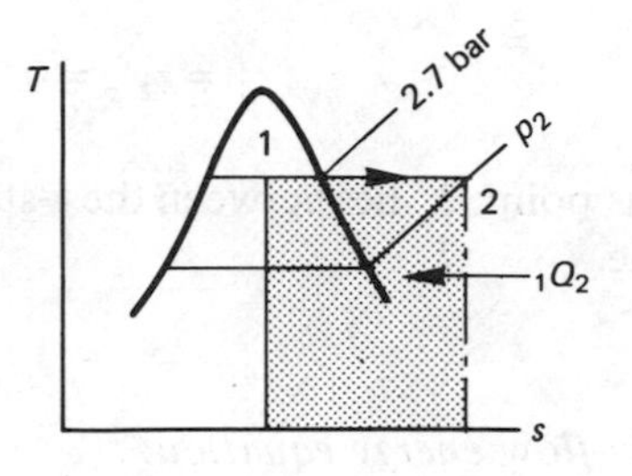

$$T_{\text{sat},1} = T_{\text{constant}} = T_2 = 130\,°\text{C}.$$

$$_1q_2 = \int_1^2 T\,\text{d}s = T_1\,(s_2 - s_1).$$

$$s_1 = 1.634 + 0.5\,(5.393) = 4.3305 \frac{\text{kJ}}{\text{kg}}$$

and $$s_2 - s_1 = \frac{_1q_2}{T_1} = \frac{1282.3 \frac{\text{kJ}}{\text{kg}}}{(130 + 273)\text{ K}} = 3.1819 \frac{\text{kJ}}{\text{K kg}}.$$

Thus $$s_2 = 4.3305 + 3.1819 = 7.5124 \frac{\text{kJ}}{\text{K kg}}$$

and, interpolating at 130 °C (p. 6 of tables),

$$p_2 = 1\text{ bar}$$

since $s_{1\text{ bar},130\,°\text{C}} = 7.36 + 0.6\,(7.614 - 7.36) = 7.5124 \frac{\text{kJ}}{\text{K kg}}$. (p. 6)

(Note that this is clearly a stage-managed result to produce 1 bar as the answer. Normally a real calculation would require a double interpolation and considerable tedium to find the final pressure.)

Example 9.7

A mass flow rate of 2 kg/s of Freon-12 flows steadily through a compressor entering at 1.509 bar, saturated vapour, and is compressed adiabatically and irreversibly with an isentropic efficiency of 0.92 to 4.914 bar.

Calculate the final temperature and the power required for compression ignoring kinetic energy.

Answer

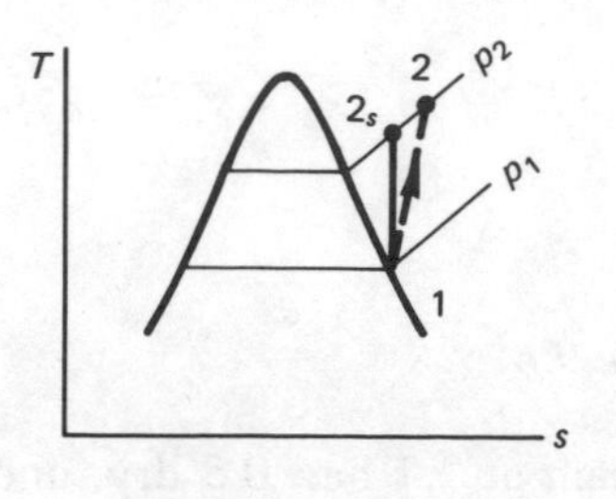

$$h_1 = 178.73 \frac{\text{kJ}}{\text{kg}}. \quad \text{(p. 13 of tables)}$$

$$s_1 = s_{2,s} = 0.7087 \frac{\text{kJ}}{\text{K kg}}.$$

Thus point 2_s lies between the g-state and the 15 K superheat state on the 4.914 bar line.

Steady-flow energy equation:

$$-{}_1\dot{W}_2 = \dot{m}(h_2 - h_1) \qquad \text{(all else is zero)}$$

or $${}_1\dot{W}_2 = \dot{m}(h_1 - h_2).$$

Now $$h_{2,s} = 193.78 + \left(\frac{0.7087 - 0.6901}{0.7251 - 0.6901}\right)(204.1 - 193.78) = 199.26 \frac{\text{kJ}}{\text{kg}}.$$

Thus $h_{2,s} - h_1 = 199.26 - 178.73 = 20.53 \dfrac{\text{kJ}}{\text{kg}}$

and $$h_2 - h_1 = \frac{h_{2,s} - h_1}{\eta} = \frac{20.53}{0.92} = 22.32 \frac{\text{kJ}}{\text{kg}}.$$

Thus $$h_2 = 178.73 + 22.32 = 201.05 \frac{\text{kJ}}{\text{kg}}$$

and $$\text{degree of superheat at 2} = \left(\frac{201.05 - 193.78}{204.10 - 193.78}\right) \times 15 \text{ K}$$

$$\text{degree of superheat at 2} = 10.57 \text{ K}.$$

Thus $$T_2 = 15\,°\text{C} + 10.57 \text{ K} = 25.57\,°\text{C}$$

and $${}_1\dot{W}_2 = 2 \frac{\text{kg}}{\text{s}} (-22.32) \frac{\text{kJ}}{\text{kg}} = -44.64 \text{ kW (work IN)}.$$

***Example** 9.8*

Ammonia (NH_3) at 10.99 bar, dryness 0.9, expands reversibly according to the law $pv^{1.06}$ = constant in an engine cylinder to a pressure of 2.908 bar. Calculate the change in specific entropy, the final dryness and the work transfer per unit mass and sketch the process on a temperature–entropy diagram.

Answer

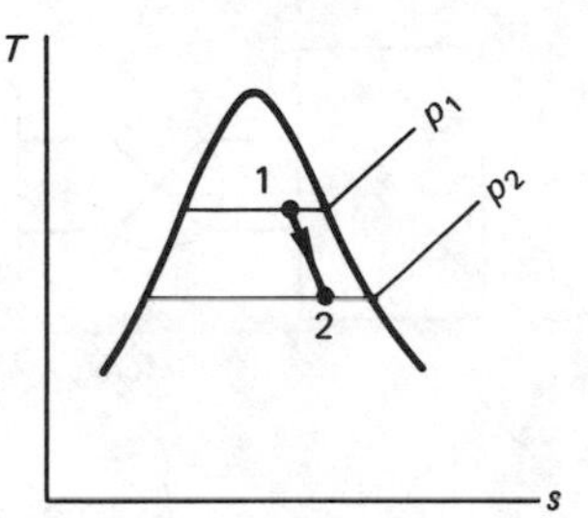

$s_1 = s_{f,1} + x_1 (s_{g,1} - s_{f,1})$ (s_{fg} not listed!)

$$= 1.171 + 0.9 (5.005 - 1.172) \quad \text{(p. 12 of tables)} \quad = 4.6207 \ \frac{\text{kJ}}{\text{K kg}} .$$

$v_1 \simeq x_1 v_{g,1} = 0.9 \times 0.1173 = 0.10557 \text{ m}^3/\text{kg}.$

$$\text{Thus} \quad v_2 = \left(\frac{p_1}{p_2}\right)^{1/n} \times v_1 = \left(\frac{10.99}{2.908}\right)^{1/1.06} \times 0.105\,57 \text{ m}^3/\text{kg} = 0.3703 \ \frac{\text{m}^3}{\text{kg}}$$

$$\text{and} \quad x_2 = \frac{v_2}{v_{g,2}} = \frac{0.3703}{0.4185} = 0.885.$$

$$s_2 = s_{f,2} + x_2 (s_g - s_f)_2 = 0.544 + 0.885 (5.475 - 0.544)$$

$$s_2 = 4.907 \ \frac{\text{kJ}}{\text{K kg}}$$

$$\text{and} \quad s_2 - s_1 = 4.907 - 4.6207 = 0.2863 \ \frac{\text{kJ}}{\text{K kg}} .$$

Specific work transfer =

$$\int_1^2 p \, \mathrm{d}v = \frac{p_1 v_1 - p_2 v_2}{n - 1} = {}_1w_2$$

$$= \frac{1099 \ \frac{\text{kN}}{\text{m}^2} \left(0.105\,57 \ \frac{\text{m}^3}{\text{kg}}\right) - 290.8 \ \frac{\text{kN}}{\text{m}^2} \left(0.3703 \ \frac{\text{m}^3}{\text{kg}}\right)}{0.06},$$

$${}_1w_2 = 138.9 \ \frac{\text{kJ}}{\text{kg}}.$$

***Example** 9.9*

2 insulated vessels, one of volume 0.1 m^3 and the other of volume 0.05 m^3, are connected by a pipe containing a closed valve. In the larger vessel there is H_2O at 3 bar, 150 °C, and the smaller vessel contains H_2O at 2 bar, 0.95 dry.

The valve is opened and the two fluids mix completely. Assuming no heat transfer with the environment, calculate the final state of the H_2O and the change of entropy.

Answer

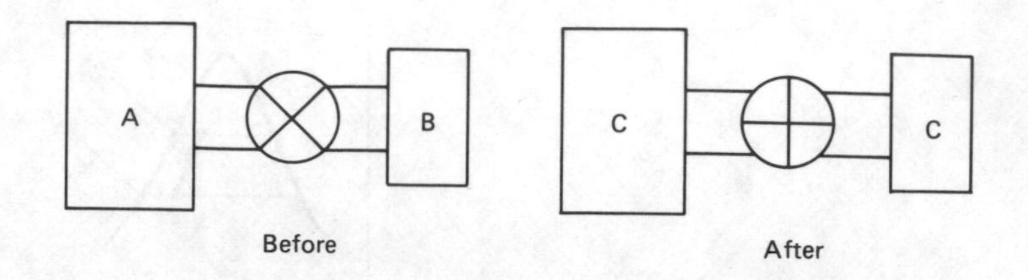

$$v_A = 0.6342 \text{ m}^3/\text{kg}. \quad \text{(p. 6 of tables)}$$

Thus
$$m_A = \frac{V_A}{v_A} = \frac{0.1 \text{ m}^3}{0.6342 \text{ m}^3/\text{kg}} = 0.1577 \text{ kg};$$

$$v_B = x_B v_{g,B} = 0.95\ (0.8856 \text{ m}^3/\text{kg}) = 0.8413 \text{ m}^3/\text{kg}; \quad \text{(p. 4)}$$

$$m_B = \frac{V_B}{v_B} = \frac{0.05 \text{ m}^3}{0.8413 \text{ m}^3/\text{kg}} = 0.0594 \text{ kg};$$

$$m_A + m_B = 0.2171 \text{ kg} = m_C$$

Energy equation (non-flow process):

$$(m_A e_A) + (m_B e_B) = (m_A + m_B) e_C.$$

$$e_A = 2572 \text{ kJ/kg}; \quad \text{(p. 6)}$$

$$e_B = 505 + 0.95\ (2530 - 505) = 2428.8 \text{ kJ/kg}. \quad \text{(p. 4)}$$

Thus

$$e_C = \frac{m_A e_A + m_B e_B}{m_A + m_B} = \frac{0.1577 \text{ kg}\ (2572 \text{ kJ/kg}) + 0.0594 \text{ kg}\ (2428.8 \text{ kJ/kg})}{0.2171 \text{ kg}}$$

$$e_C = 2532.8 \text{ kJ/kg}.$$

Also
$$V_C = 0.1 + 0.05 = 0.15 \text{ m}^3.$$

Thus
$$v_C = \frac{V_C}{m_C} = \frac{0.15 \text{ m}^3}{0.2171 \text{ kg}} = 0.6909 \text{ m}^3/\text{kg}.$$

The values of e_C and v_C give the state required, and, interpolating in the tables,

$$p_C = 2.6 \text{ bar}, \qquad x_C = 0.997 \text{ dry}.$$

$\Delta S_{A \to C} = m_A (s_C - s_A)$ for the fluid in A changing to C,

$\Delta S_{B \to C} = m_B (s_C - s_B)$ for the fluid in B changing to C.

Now $s_A = 7.078 \dfrac{\text{kJ}}{\text{K kg}}$ (p. 6); $s_B = 1.530 + 0.95\ (5.597) = 6.847 \dfrac{\text{kJ}}{\text{K kg}}$; (p. 4)

$$s_C = 1.621 + 0.997\ (5.419) = 7.024 \frac{\text{kJ}}{\text{K kg}} \quad \text{(p. 4)}.$$

Thus
$$\Delta S_{(A+B) \to C} = 0.1577 \text{ kg}\ (7.024 - 7.078) \frac{\text{kJ}}{\text{K kg}} + 0.0594 \text{ kg}\ (7.024 - 6.847) \frac{\text{kJ}}{\text{K k}}$$

$$\Delta S_{(A+B) \to C} = 0.001\,998 \frac{\text{kJ}}{\text{K}}.$$

9.3 Gases and Gas Mixtures—Entropy Changes

The calculation of entropy changes in the case of perfect gases and gas mixtures is a matter of simple substitution into equation 4 (page 90).

The most useful form of equation 4 is

$$s_2 - s_1 = c_p \ln \frac{T_2}{T_1} - R \ln \frac{p_2}{p_1} \qquad \text{since } p \text{ and } T \text{ are easily measured.}$$

Alternatively,

$$s_2 - s_1 = c_v \ln \frac{T_2}{T_1} + R \ln \frac{v_2}{v_1}$$

and additionally we derive as follows:

$$s_2 - s_1 = c_v \ln \frac{T_2}{T_1} + (c_p - c_v) \ln \frac{v_2}{v_1} \qquad (R = c_p - c_v)$$

$$= c_p \ln \frac{v_2}{v_1} + c_v \left[\ln \frac{T_2}{T_1} - \ln \frac{v_2}{v_1}\right]$$

$$= c_p \ln \frac{v_2}{v_1} + c_v \left[\ln \frac{T_2 v_1}{T_1 v_2}\right]$$

$$= c_p \ln \frac{v_2}{v_1} + c_v \ln \frac{p_2}{p_1}.$$

For a mixture of gases changing state, the entropy change will depend on the values of $c_{p,t}$, $c_{v,t}$ and R_t for the mixture. For single gases combining to form a mixture you will need to treat each gas separately as in the last part of example 9.9 (see page 97). The inference here is that each individual gas retains its own identity, even on mixing, which is the assumption underlying the theory of gas mixtures.

9.4 Worked Examples

Example 9.10

One kg of air at 1 bar, 20 °C, is compressed according to the law $pv^{1.3}$ = constant until the pressure is 5 bar.

Calculate the change in entropy and sketch the process on a T–s diagram indicating the area representing heat flow.

Answer

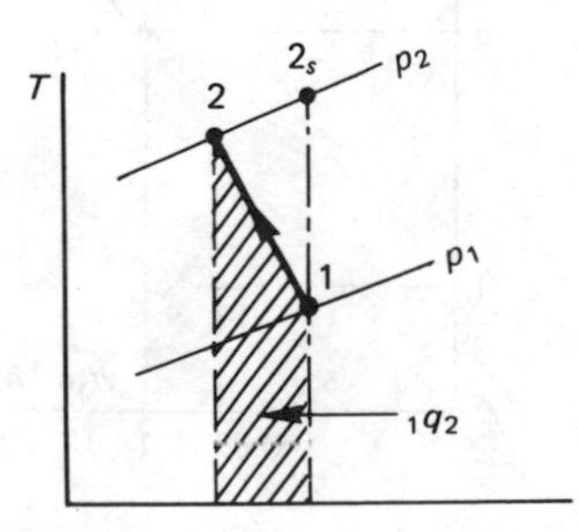

For isentropic compression ($\gamma_{air} = 1.4$)

$$T_{2,s} = T_1 \left(\frac{p_2}{p_1}\right)^{(\gamma-1)/\gamma} = 293 \text{ K } (5)^{0.286} = 464.3 \text{ K}.$$

$$\text{Actual} \qquad T_2 = T_1 \left(\frac{p_2}{p_1}\right)^{(n-1)/n} = 293\ \text{K}\ (5)^{0.231} = 424.94\ \text{K}.$$

$$\text{Now} \qquad s_2 - s_1 = s_2 - s_{2,s} \qquad \text{(along same isobar } p_2\text{)}$$

$$\text{and} \qquad s_2 - s_{2,s} = s_2 - s_1 = c_p \ln \frac{T_2}{T_{2,s}} \qquad \left(\text{since } R \ln \frac{p_2}{p_{2,s}} = 0\right)$$

$$s_2 - s_1 = 1.005\ \frac{\text{kJ}}{\text{kg K}} \ln \frac{424.94}{464.3} = -0.089\ \frac{\text{kJ}}{\text{K kg}}.$$

Example 9.11

1 kg of air at 1 bar, 25 °C, changes its state to 6 bar and a volume of 1 m³. Calculate the change of entropy and sketch the initial and final state points on the p–v and T–s fields.

Answer

$$\text{Per kg fluid,} \quad v_1 = \frac{RT_1}{p_1} = \frac{0.287\ \frac{\text{kJ}}{\text{kg K}} \times 298\ \text{K}}{100\ \frac{\text{kN}}{\text{m}^2}} = 0.8553\ \frac{\text{m}^3}{\text{kg}},$$

$$v_2 = 1\ \frac{\text{m}^3}{\text{kg}},$$

$$T_2 = \frac{p_2 v_2}{R} = \frac{600\ \frac{\text{kN}}{\text{m}^2} \times 1\ \frac{\text{m}^3}{\text{kg}}}{0.287\ \frac{\text{kJ}}{\text{kg K}}} = 2090.6\ \text{K}.$$

$$s_2 - s_1 = c_p \ln \frac{v_2}{v_1} + c_v \ln \frac{p_2}{p_1}$$

$$= 1.005\ \frac{\text{kJ}}{\text{kg K}} \ln \frac{1}{0.8853} + 0.718\ \frac{\text{kJ}}{\text{kg K}} \ln 6$$

$$= 0.1224 + 1.2865 = 1.4089\ \frac{\text{kJ}}{\text{K kg}}.$$

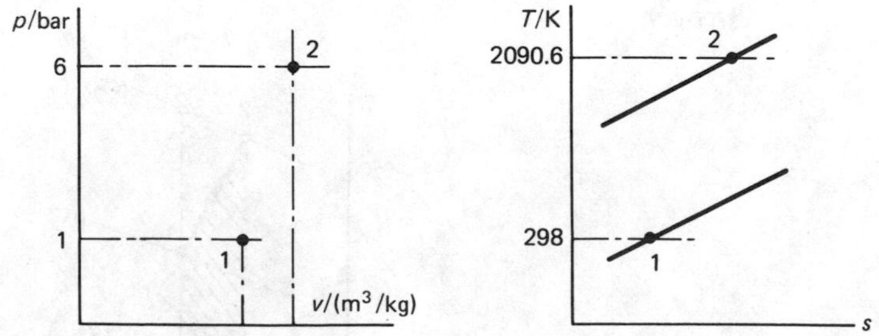

Note that $(s_2 - s_1)$ can be calculated even though the mechanism of the state change is unknown This is because both the initial and final states are known and entropy is a *property* whose value depends solely upon the state values and is independent of the path taken between them even when the process is irreversible.

Example 9.12

0.5 m^3 ethane (C_2H_4) at 7 bar, 260 °C expands isentropically in a cylinder behind a piston to 1 bar, 100 °C.

Calculate the work done in expansion assuming that ethane is a perfect gas. (c_p = 1.89 kJ/(kg K).)

The same mass is now recompressed back to 7 bar according to the law $pv^{1.35}$ = constant. Calculate the final temperature and the heat transfer. Calculate also the change in entropy and sketch both processes on the p–v and T–s fields.

Answer

$$R(C_2H_4) = \frac{R_0}{m_v} = \frac{8.3143 \dfrac{\text{kJ}}{\text{kmol K}}}{28 \dfrac{\text{kg}}{\text{kmol}}} = 0.297 \frac{\text{kJ}}{\text{kg K}};$$

$$c_v(C_2H_4) = c_p - R = 1.89 - 0.297 = 1.593 \frac{\text{kJ}}{\text{kg K}};$$

$$\gamma(C_2H_4) = \frac{c_p}{c_v} = \frac{1.89}{1.593} = 1.186.$$

$$m = \left(\frac{pV}{RT}\right)_1 = \frac{700 \dfrac{\text{kN}}{\text{m}^2} \times 0.5 \text{ m}^3}{0.297 \dfrac{\text{kJ}}{\text{kg K}} \times 533 \text{ K}} = 2.211 \text{ kg}$$

and

$$_1W_2 = \frac{mR(T_1 - T_2)}{\gamma - 1} = \frac{2.211 \text{ kg} \times 0.297 \dfrac{\text{kJ}}{\text{kg K}} (260 - 100) \text{ K}}{0.186} = 564.9 \text{ kJ}.$$

$$T_3 = T_2 \left(\frac{p_3}{p_2}\right)^{(n-1)/n} = 373 \text{ K} \left(\frac{7}{1}\right)^{0.259} = 617.43 \text{ K} \qquad (n = 1.35).$$

Energy equation:

$$_2Q_3 - {}_2W_3 = E_3 - E_2 = mc_v(T_3 - T_2);$$

$$_2Q_3 = {}_2W_3 + mc_v(T_3 - T_2) = \frac{mR(T_2 - T_3)}{n - 1} + \frac{mR}{\gamma - 1}(T_3 - T_2)$$

$$= \frac{mR(T_2 - T_3)}{(n - 1)} \left(\frac{\gamma - n}{\gamma - 1}\right) = \left(\frac{\gamma - n}{\gamma - 1}\right) {}_2W_3 \qquad \text{(a standard result);}$$

$$_2W_3 = \frac{2.211 \text{ kg} \times 0.297 \dfrac{\text{kJ}}{\text{kg K}} (373 - 617.43) \text{ K}}{0.35} = -458.6 \text{ kJ}.$$

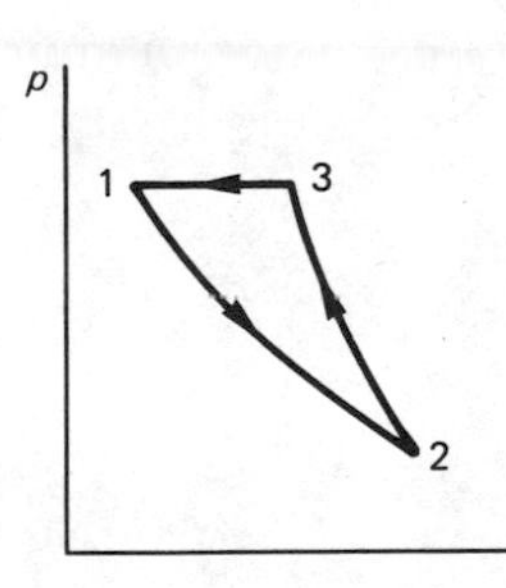

(a)

Thus $$_2Q_3 = \frac{1.186 - 1.35}{0.186} (-458.6) \text{ kJ} = +404.4 \text{ kJ} \quad \text{(added)}$$

$$S_3 - S_2 = mc_p \ln \frac{T_3}{T_2} - mR \ln \frac{p_3}{p_2}$$

$$= 2.211 \text{ kg} \left[1.89 \frac{\text{kJ}}{\text{kg K}} \ln \frac{617.7}{373} - 0.297 \frac{\text{kJ}}{\text{kg K}} \ln 7\right]$$

$$= 2.211 \text{ kg} [0.9533 - 0.5779] \frac{\text{kJ}}{\text{K kg}} = 0.83 \frac{\text{kJ}}{\text{K}}.$$

T
s
1
2
3
(b)

Example 9.13

Air flows through a perfectly insulated duct. At one section A the pressure and temperature are respectively 2 bar and 200 °C and at another section B further along the duct the corresponding values are 1.5 bar and 150 °C.

Which way is the air flowing?

This is a good example to show the value of entropy.

This question cannot be solved by application of the first law of thermodynamics since there is nothing to tell us whether the fluid is expanding from A to B or being compressed from B to A.

However, since the duct is insulated the inference is that there is no heat transfer to or from the environment and therefoe there is no change of entropy in the environment.

Now in any real process the change of entropy between the system and the surroundings must be positive.

Thus by calculating the entropy change for the system we can quickly discover the sense of the flow since the entropy must increase in the same sense.

Try $$s_B - s_A = c_p \ln \frac{T_B}{T_A} - R \ln \frac{p_B}{p_A}$$

$$= 1.005 \frac{\text{kJ}}{\text{kg K}} \ln\left(\frac{273 + 150}{273 + 200}\right) - 0.287 \frac{\text{kJ}}{\text{kg K}} \ln \frac{1.5}{2}$$

$$= -0.112\,26 + 0.0826 = -0.029\,66 \frac{\text{kg}}{\text{K kg}} .$$

Thus $s_A > s_B$ and the flow is from B to A.

So that even though entropy cannot be measured directly it can still be used to find the sense of a flow in a well-insulated duct given two salient states as above.

Example 9.14

2 kg of ammonia (NH_3) are throttled from 4.295 bar, 50 °C to 2.077 bar. Calculate the change of entropy. Sketch the process on an h–s field.

Answer

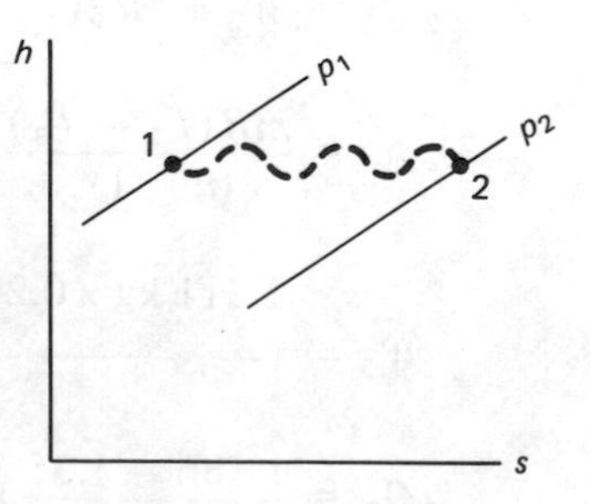

(Since an irreversible process is involved only end states 1 and 2 can be depicted.)

From page 12 of tables, at 4.295 bar,

$$T_{sat} = 0\ °\text{C}.$$

But $$T_1 = 50\ °\text{C}.$$

Thus degree of superheat at 1 = 50 K.

Thus

$$h_1 = 1567.8 \ \frac{\text{kJ}}{\text{kg}} = h_2 \qquad \text{(from the energy equation all other terms are zero).}$$

Now at 2.077 bar, $\quad h_2 = 1567.8 \ \dfrac{\text{kJ}}{\text{kg}}$.

$$s_2 = 6.008 + \left(\frac{1567.8 - 1538.2}{1650.0 - 1538.2}\right)(6.347 - 6.008) = 6.0978 \ \frac{\text{kJ}}{\text{K kg}}.$$

$$s_2 - s_1 = 0.341\,75 \ \frac{\text{kJ}}{\text{K kg}}.$$

And

$$S_2 - S_1 = m(s_2 - s_1) = 2 \times 0.341\,75 = 0.6835 \ \frac{\text{kJ}}{\text{K}}.$$

Example 9.15

A rigid insulated vessel of 1 m^3 capacity contains air at 2 bar, 50 °C. It is isolated from a second rigid insulated vessel of capacity 0.5 m^3 containing CO_2 at 1 bar, 40 °C, by a pipe containing a valve. The valve is opened and the two fluids mix freely. Find the final state when equilibrium is finally re-established and also the change in entropy. (c_p for CO_2 is 0.871 kJ/(kg K).)

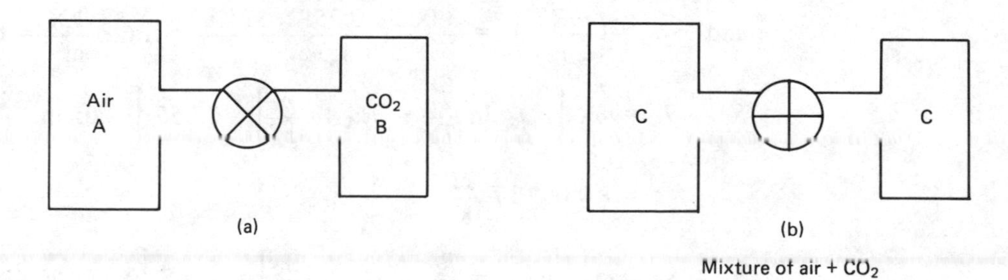

Answer

Energy equation (non-flow process with boundary round both vessels):

$$m_A e_A + m_B e_B = (m_A + m_B)(e_C); \qquad \text{(all other terms zero)}$$

$$m_A = \left(\frac{pV}{RT}\right)_A = \frac{200 \ \frac{\text{kN}}{\text{m}^2} \times 1 \ \text{m}^3}{0.287 \ \frac{\text{kJ}}{\text{kg K}} \times 323 \ \text{K}} = 2.157 \ \text{kg},$$

$$m_B = \left(\frac{pV}{RT}\right)_B = \frac{100 \times 0.5}{\frac{8.3143}{44} \times 313} = 0.845 \ \text{kg}.$$

Thus $\quad m_C = m_A + m_B = 3.002$ kg.

$$c_v(CO_2) = c_p - R = 0.871 \ \frac{\text{kJ}}{\text{kg K}} - \frac{8.3143 \ \frac{\text{kJ}}{\text{kmol K}}}{44 \ \frac{\text{kJ}}{\text{kmol}}}$$

$$c_v(CO_2) = 0.871 - 0.189 = 0.682 \ \frac{\text{kJ}}{\text{kg K}}.$$

$$\text{Thus } e_C = \frac{m_A e_A + m_B e_B}{m_A + m_B} = \frac{m_A c_{v,A} T_A + m_B c_{v,B} T_B}{m_A + m_B}$$

$$= \frac{\left(2.157 \text{ kg} \times 0.718 \ \frac{\text{kJ}}{\text{kg K}} \times 323 \text{ K}\right) + \left(0.845 \text{ kg} \times 0.682 \ \frac{\text{kJ}}{\text{kg K}} \times 313 \text{ K}\right)}{3.002 \text{ kg}}$$

$$= 226.7 \ \frac{\text{kJ}}{\text{kg}},$$

and

$$V_C = V_A + V_B = 1.5 \text{ m}^3$$

or

$$v_C = \frac{V_C}{m_C} = \frac{1.5 \text{ m}^3}{3.002 \text{ kg}} = 0.5 \ \frac{\text{m}^3}{\text{kg}}$$

$$\text{Now } R_C = \frac{\Sigma_i mR}{\Sigma_i m} = \frac{(2.157 \times 0.287) + (0.845 \times 0.189)}{3.002} = 0.259 \ \frac{\text{kJ}}{\text{kg K}}$$

$$\text{and } c_{v,C} = \frac{\Sigma_i m c_v}{\Sigma_i m} = \frac{(2.157 \times 0.718) + (0.845 \times 0.682)}{3.002} = 0.708 \ \frac{\text{kJ}}{\text{kg K}}.$$

$$\text{Thus } T_C = \frac{e_C}{c_{v,C}} = \frac{227.8 \ \frac{\text{kJ}}{\text{kg}}}{0.708 \ \frac{\text{kJ}}{\text{kg K}}} = 321.8 \text{ K } (48.8\ °\text{C})$$

$$\text{and } p_C = \left(\frac{mRT}{V}\right)_C = \frac{3.002 \times 0.259 \times 321.8}{1.5} = 166.8 \ \frac{\text{kN}}{\text{m}^2} = 1.668 \text{ bar.}$$

$$S_C - S_A = m_A \left[c_{p,A} \ln \frac{T_C}{T_A} - R_A \ln \frac{p_C}{p_A}\right] = 2.157 \left[1.005 \ln \frac{320.3}{323} - 0.287 \ln \frac{1.66}{2}\right]$$

$$= +0.097\,17 \ \frac{\text{kJ}}{\text{K}}.$$

$$S_C - S_B = m_B \left[c_{p,B} \ln \frac{T_C}{T_B} - R_B \ln \frac{p_C}{p_B}\right] = 0.845 \left[0.871 \ln \frac{320.3}{313} - 0.259 \ln \frac{1.66}{1}\right]$$

$$= -0.093\,95 \ \frac{\text{kJ}}{\text{K}}.$$

$$\therefore \Delta S_{(A+B)\rightarrow C} = 0.097\,17 - 0.093\,95 = 0.003\,22 \text{ kJ/K}.$$

Example 9.16

A rigid cylinder whose volume is 0.1 m^3 is divided into two compartments by a freely sliding (frictionless) non-conducting piston of negligible volume. Initially one compartment, A, has a volume of 0.09 m^3 and contains steam at 3.5 bar, dryness 0.85. The other compartment, B, contains steam also at 3.5 bar but 0.15 dry. Compartment A is insulated against heat transfer with the surroundings. The steam in compartment B is heated slowly until the pressure is 7 bar.

Determine the final condition of the steam in each compartment.

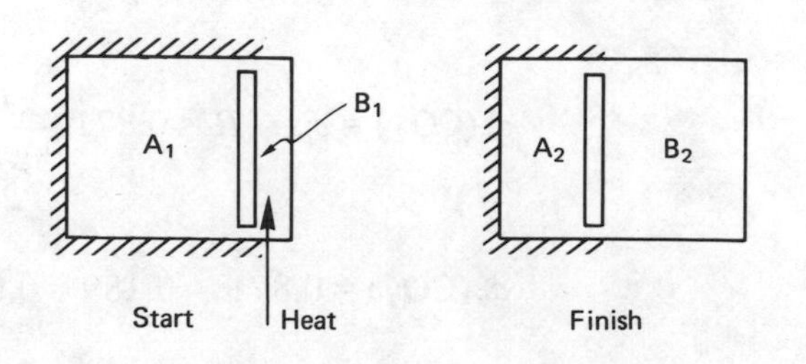

Answer

Once heating is over and equilibrium is established, the piston will be stationary; i.e.

$$p_{A,2} = p_{B,2} = 7 \text{ bar.}$$

$$m_A = \frac{V_{A,1}}{v_{A,1}} = \frac{V_{A,1}}{x_{A,1}v_{g,A,1}} \quad \text{(ignoring } v_f\text{)}$$

$$= \frac{0.09 \text{ m}^3}{0.85 \times 0.5241 \dfrac{\text{m}^3}{\text{kg}}} = 0.202 \text{ kg.} \quad \text{(p. 4 of tables at 3.5 bar)}$$

$$s_{A,1} = s_{f,A,1} + x_{A,1}s_{fg,A,1} = 1.727 + 0.85\,(5.214) = 6.1589 \text{ kJ/K kg.}$$

$$x_{A,2} = \frac{s_{A,2} - s_{f,A,2}}{s_{fg,A,2}} = \frac{6.1589 - 1.992}{4.717} = 0.8834. \quad \text{(p. 4 at 7 bar)}$$

(since the compartment is insulated and the piston moves without friction), i.e.

$$s_{A,1} = s_{A,2}.$$

$$m_{B,1} = m_{B,2} = \frac{V_{B,1}}{v_{B,1}} = \frac{V_{B,1}}{x_{B,1}v_{g,B,1}} = \frac{0.01 \text{ m}}{0.15 \times 0.5241 \text{ m/kg}} = 0.1272 \text{ kg.}$$

$$V_{A,2} = m_{A,2}v_{A,2} = m_{A,2}x_{A,2}v_{g,A,2} = 0.202 \text{ kg} \times 0.8834 \times 0.2728 \text{ m}^3/\text{kg}$$

$$= 0.0487 \text{ m}^3.$$

$$V_{B,2} = 0.1 - 0.0487 \text{ m}^3 = 0.0513 \text{ m}^3.$$

$$v_{B,2} = \frac{V_{B,2}}{m_B} = \frac{0.0513 \text{ m}^3}{0.1272 \text{ kg}} = 0.4033 \frac{\text{m}^3}{\text{kg}} \text{ at 7 bar.}$$

Thus from tables page 7 at 7 bar:

$$T_{B,2} = 300\,^\circ\text{C} + \left(\frac{0.4035 - 0.3714}{0.4058 - 0.3714}\right)(50 \text{ K}) = 346.5\,^\circ\text{C}.$$

Exercises

1 Steam at 20 bar is throttled to 1 bar, 150 °C. Calculate the initial steam quality and the change in specific entropy, making the usual assumptions.
(0.988, –1.3193 kJ/(K kg))

2 Steam expands adiabatically but irreversibly from 30 bar, 350 °C in a turbine to 1 bar, 0.914 dry. Calculate the ratio of the actual work transfer to the reversible work transfer (or isentropic efficiency), neglecting both kinetic and potential energies and also find the change in speicific entropy. (0.87, +0.2549 kJ/(K kg))

3 Two vessels whose volume ratio is two are connected by a valve and kept at constant temperature by immersion in a large reservoir of water of fixed temperature. The smaller vessel initially contains CO_2 and the larger one is evacuated.

Calculate the change in specific entropy when the valve is opened and conditions settle. Assume CO_2 is a perfect gas. (+0.2083 kJ/(K kg))

4 Steam is expanded isothermally at 200 °C from 10 bar to 6 bar and then further expanded isentropically to 1 bar. Calculate the overall change in specific entropy, the total heat transfer and the total work transfer stating the sense in each case. (0.273 kJ/(K kg), +129.1 kJ/kg, +427.1 kJ/kg).

5 Nitrogen is compressed adiabatically through a pressure ratio of 5 in a rotary compressor. The ratio of final to initial temperatures is 1.68 and the ratio of specific heat capacities is 1.4.

Show that the process is irreversible and calculate the change in specific entropy. (+0.0612 kJ/(K kg))

6 It is possible, conceptually, to conceive of an irreversible process whose overall change of entropy is zero provided the heat transfer is adjusted to suit.

Calculate the specific heat transfer stating the sense for an irreversible non-flow compression process for air between an initial state of 1 bar, 20 °C, and a final pressure of 5 bar with a work input of 100 kJ/kg with no overall change of specific entropy and neglecting k.e. and p.e. (+37.1 kJ/kg)

10 Vapour Power Cycles

10.1 Introduction

The ideal cycle on which practical vapour power plants are based is the Rankine cycle.

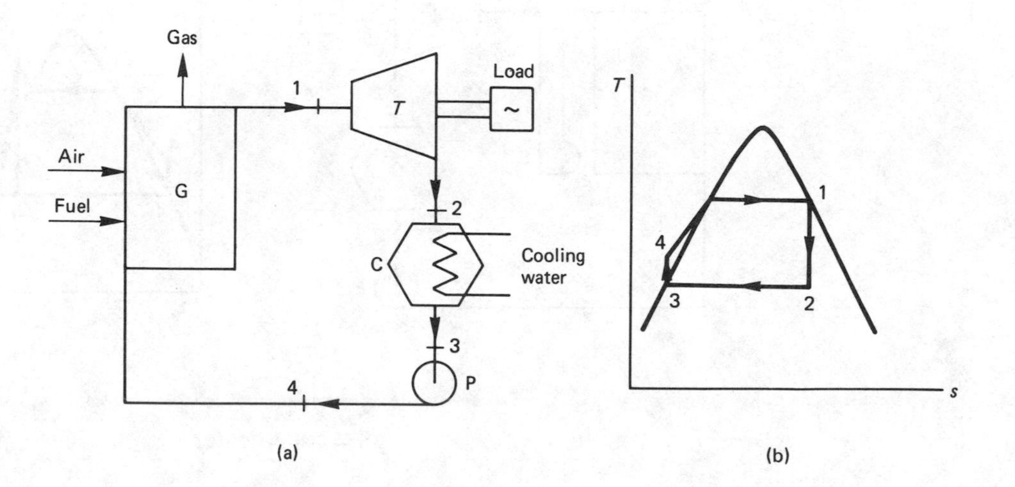

G ≡ steam generator; T ≡ turbine; C ≡ condenser; P ≡ feed pump

In the above schematic the following assumptions are made:

(a) The turbine expansion is isentropic.
(b) The turbine inlet steam is just saturated vapour.
(c) The same fluid recirculates continuously.
(d) The feed pump work is small but can be significant.

This cycle gives a classic example of the use of entropy. Without entropy state 2 cannot be determined, as normally only the pressure at 2 is specified.

$_1w_2 = h_1 - h_2$ (positive work out, from SFEE ignoring k.e. and p.e.).

$_3w_4 = h_3 - h_4$ (negative work in to feed pump).

$_4q_1 = h_1 - h_4$ (positive energy added to cycle by fuel in G).

$_2q_3 = h_2 - h_3$ (negative energy rejected to cooling water in C).

$$\eta_{\text{thermal}} = \frac{_1w_2 + {_3w_4}}{_4q_1} = \frac{(h_1 - h_2) + (h_3 - h_4)}{h_1 - h_4} = \frac{(h_1 - h_2) + (h_3 - h_4)}{(h_1 - h_3) + (h_3 - h_4)};$$

$$\eta_{\text{th}} \text{ (approx.)} = \frac{_1w_2}{_3q_1} \quad \text{(ignoring feed pump work)}$$

The above shows the four successive applications of the steady-flow energy equation to each of the four components in the plant, T, P, G and C respectively.

Practical limitations:

(a) The turbine exhaust must be reasonably dry to avoid blade erosion ($x_2 \nless 0.9$ approximately).

(b) Expansion is not isentropic but can be assumed adiabatic ($\eta_T = 0.8 - 0.9$ approximately).

(c) The turbine expansion begins with steam superheated in order that the exit quality of steam does not fall below about 0.9 to avoid blade erosion problems.

The modified Rankine cycle with superheated steam at inlet to turbine and irreversible expansion is shown here.

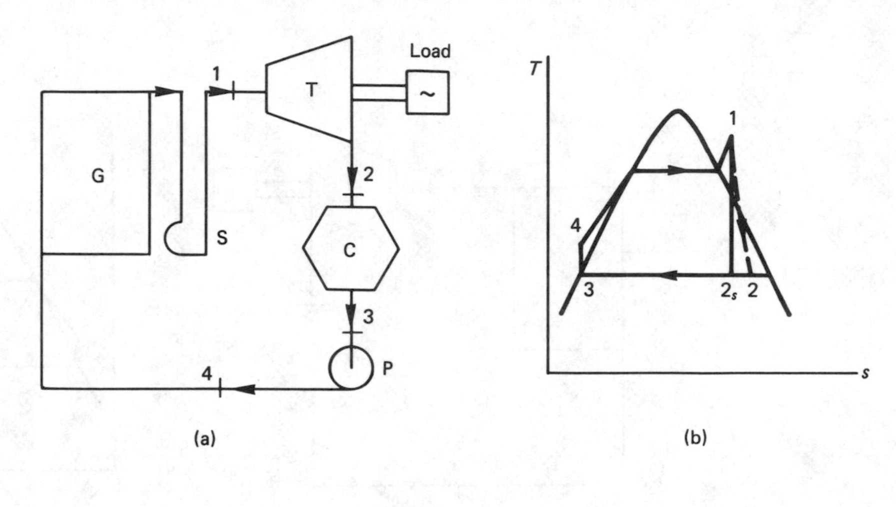

S ≡ superheater

Further improvements:
In a modern vapour power plant it is necessary to increase capital expenditure in order to improve running costs:

(a) by reheating of the turbine steam after partial expansion,

(b) by regenerative feed heating of the water on return to the steam generator by means of partial bleeding of the steam from the turbine expansion.

It is quite normal to have as many as nine stages of feed heating with the fluid cascading down the pressure scale from one heater to the next and finally to the condenser.

Criteria of performance:

(a) Thermal efficiency $\eta_{th} = \dfrac{\text{Net work transfer in cycle}}{\text{Positive heat supplied}}$.

(b) Work ratio $r_w = \dfrac{\text{Net work transfer}}{\text{Gross work transfer}}$

(nearly unity for the Rankine cycle since feed pump work is very small).

(c) Specific steam consumption $= \dfrac{\text{Mass flow rate}}{\text{Power output}}$

(a *size* criterion).

10.2 Worked Examples

Example 10.1

Saturated liquid (H_2O) is fed into a boiler, evaporated at a constant temperature of 212.4 °C, expanded in an adiabatic turbine and condensed at a constant temperature of 93.5 °C. The saturated liquid leaving the condenser is then compressed, heated at constant pressure and returned to the boiler as saturated liquid at 212.4 °C.

The energy rejected in the condenser is 1870 kJ/kg and the mean specific heat capacity of the liquid over the temperature range 93.5 °C to 212.4 °C is 4.33 kJ/(kg K).

Determine without using property tables the ratio

$$\frac{\text{Enthalpy change in turbine}}{\text{Enthalpy change in liquid heating + Enthalpy change in evaporation}}$$

which is known as the thermal efficiency of the cycle.

Assume all processes are reversible, neglect all changes in kinetic and potential energy and ignore the small difference between the compressed liquid enthalpy and the saturation enthalpy at 93.5 °C.

Answer

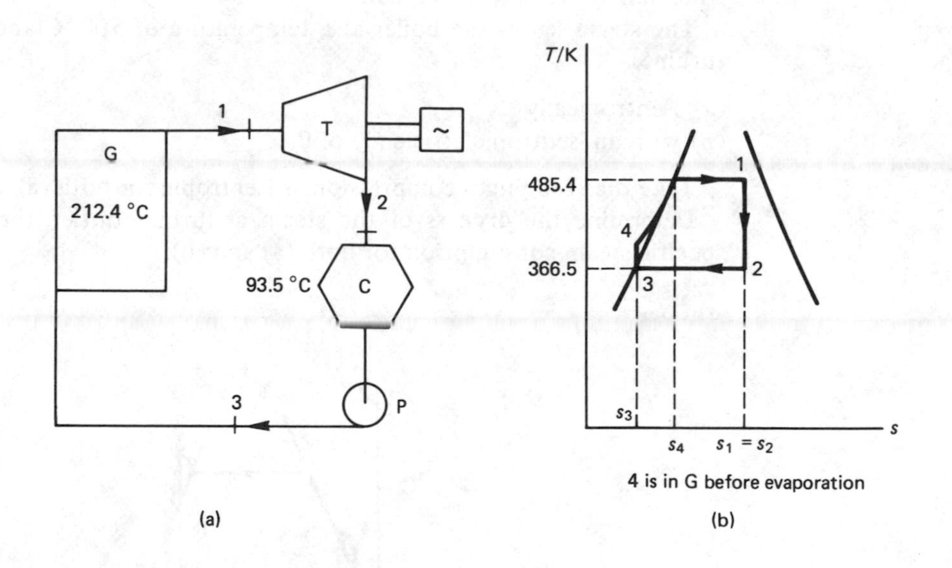

Feed pump work is to be ignored here.

$$_2Q_3 = \text{area under line 2–3 on } T\text{–}s \text{ diagram} = \int_2^3 T\,\mathrm{d}s$$

$$= T_2\,(s_3 - s_2) = 366.5\ \mathrm{K}\,(s_3 - s_2) = -\,1870\ \frac{\mathrm{kJ}}{\mathrm{kg}}\,.$$

$$s_2 - s_3 = \frac{1870}{366.5} = 5.1023\ \frac{\mathrm{kJ}}{\mathrm{K\,kg}}\,.$$

$$s_4 - s_3 = \int_3^4 \frac{dQ_R}{T} = \int_3^4 \frac{c_p\,dT}{T} = c_p \int_3^4 \frac{dT}{T} = c_p \ln \frac{T_4}{T_3}$$

$$= 4.33 \frac{\text{kJ}}{\text{kg K}} \ln \frac{485.4}{366.5} = 1.2166 \frac{\text{kJ}}{\text{K kg}}.$$

$$s_1 - s_4 = (s_2 - s_3) - (s_4 - s_3) \qquad \text{since } s_1 = s_2$$

$$= 5.1023 - 1.2166 = 3.8857 \frac{\text{kJ}}{\text{K kg}}.$$

$${}_3Q_4 = c_p\,(T_4 - T_3) = 4.33 \frac{\text{kJ}}{\text{kg K}} (485.4 - 366.5)\,\text{K} = 514.8 \frac{\text{kJ}}{\text{kg}}.$$

$${}_4Q_1 = T_4\,(s_1 - s_4) = 488.4\,\text{K} \left(3.8857 \frac{\text{kJ}}{\text{K kg}}\right) = 1897.8 \frac{\text{kJ}}{\text{kg}}.$$

or

$${}_1W_2 = {}_3Q_4 + {}_4Q_1 + {}_2Q_3 \qquad (\Sigma\, Q = 0 \text{ in a cycle})$$

$$= 514.8 + 1897.8 - 1870 = 542.6 \frac{\text{kJ}}{\text{kg}}.$$

$$\eta_{\text{th}} = \frac{{}_1W_2}{{}_3Q_4 + {}_4Q_1} = \frac{531}{514.8 + 1897.8} = 0.220.$$

Example 10.2

A steam plant operates on the Rankine cycle with a boiler pressure of 70 bar and a condenser pressure of 0.5 bar.

The steam leaves the boiler at a temperature of 500 °C and is expanded in the turbine,

(a) isentropically
(b) with an isentropic efficiency of 0.8.

Take the feed pump compression as isentropic for both (a) and (b).

Determine the dryness of the steam at turbine outlet, the efficiency and the specific steam consumption for both (a) and (b).

Answer

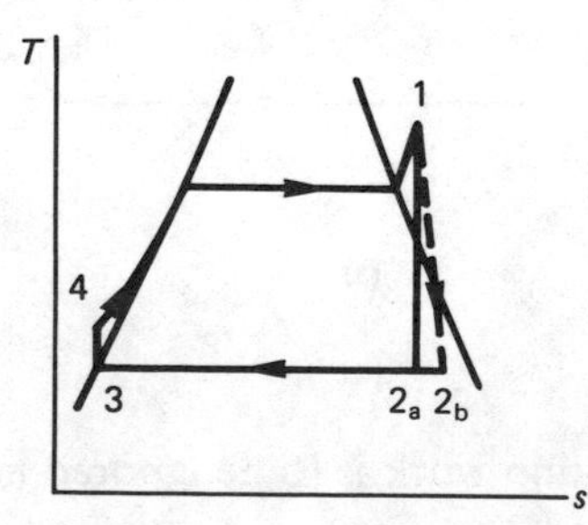

$$s_1 = 6.796 \frac{\text{kJ}}{\text{K kg}} \quad \text{(p. 7 of tables)} \quad (= s_{2,a}).$$

$$h_1 = 3410 \frac{\text{kJ}}{\text{kg}} \quad \text{(p. 7)}.$$

$$x_{2,a} = \frac{s_{2,a} - s_{f,2}}{s_{fg}} = \frac{6.796 - 1.091}{6.502} = 0.8774 \quad \text{(p. 3)}.$$

$$h_{2,a} = h_{f,2} + x_{2,a}h_{fg,2} = 340 + 0.8774(2305) = 2362.5 \ \frac{kJ}{kg}. \text{ (p. 3).}$$

$$h_1 - h_{2,a} = 1047.5 \ \frac{kJ}{kg} = w_{ideal}.$$

$$h_1 - h_{2,b} = \eta(h_1 - h_{2,a}) = 0.8(1047.5) = 838 \ \frac{kJ}{kg} = w_{irreversible}.$$

$$h_{2,b} = 3410 - 838 = 2572 \ \frac{kJ}{kg}.$$

$$x_{2,b} = \frac{2572 - 340}{2305} = \frac{h_{2,b} - h_{f,2}}{h_{fg,2}} = 0.968.$$

$$\text{Feed pump work} = -\int_3^4 v\,dp \simeq 0.001\,037 \ \frac{m^3}{kg} \ (7000 - 50) \ \frac{kN}{m^2} \quad \text{(p. 10)}$$

$$\text{Feed pump work} = h_4 - h_3 = 7.21 \ \frac{kJ}{kg}.$$

$$\eta_{ideal} = \frac{(h_1 - h_{2,a}) - (h_4 - h_3)}{(h_1 - h_3) - (h_4 - h_3)} = \frac{(h_1 - h_{2,a}) - (h_4 - h_{f,2})}{(h_1 - h_{f,2}) - (h_4 - h_{f,2})}$$

$$= \frac{1047.5 - 7.21}{3410 - 340 - 7.21} = 0.34.$$

$$\eta_{irreversible} = \frac{838 - 7.21}{3410 - 340 - 7.21} = 0.271.$$

$$\text{Specific steam consumption (ideal)} = \frac{1}{1047.5} \ \frac{kg}{kJ} \left[\frac{kJ}{s\,kW}\right] \left[\frac{3600\,s}{h}\right]$$

$$= 3.437 \ \frac{kg}{kW\,h}.$$

$$\text{Irreversible s.s.c.} = \frac{3600}{838} = 4.296 \ \frac{kg}{kW\,h}.$$

Example 10.3

The design criteria for a vapour power cycle operating on the Rankine cycle are:

Mass flow rate 4000 kg/min
Power output 50 MW
Inlet turbine pressure 40 bar
Exhaust pressure 0.1 bar
Quality at turbine outlet 0.94

Ignoring feed pump work and changes in kinetic and potential energies find

(a) the turbine inlet temperature,
(b) the thermal efficiency,
(c) the turbine isentropic efficiency,
(d) the specific steam consumption.

Answer

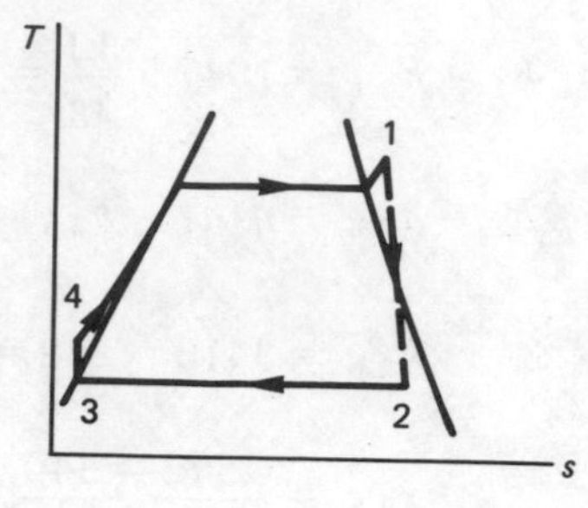

$$h_1 - h_2 = \frac{{}_1\dot{W}_2}{\dot{m}} = \frac{50\,000 \text{ kW}}{4000 \dfrac{\text{kg}}{\text{min}}} \left[\frac{60 \text{ s}}{\text{min}}\right] \left[\frac{\text{kJ}}{\text{s kW}}\right]$$

$$h_1 - h_2 = 750 \ \frac{\text{kJ}}{\text{kg}}.$$

$$h_2 = h_{f,2} + x_2 h_{fg,2} = 192 + 0.94\ (2392) = 2440.5 \ \frac{\text{kJ}}{\text{kg}}. \quad \text{(p. 3 of tables)}$$

$$h_1 = 2440.5 + 750 = 3190.5 \ \frac{\text{kJ}}{\text{kg}}.$$

$$T_1 = 350\ °\text{C} + \left(\frac{3190.5 - 3094}{3214 - 3094}\right) (50 \text{ K}) = 390.2\ °\text{C}. \quad \text{(p. 7)}$$

$${}_2\dot{Q}_3 = \dot{m}\ (h_2 - h_{f,2}) = 4000 \ \frac{\text{kg}}{\text{min}} \left[\frac{\text{min}}{60 \text{ s}}\right] 0.94 \times 2392 \ \frac{\text{kJ}}{\text{kg}} = 149\,900 \text{ kW} = 149.9 \text{ MW}.$$

$$\eta_{th} = \frac{{}_1\dot{W}_2}{{}_1\dot{W}_2 + {}_2\dot{Q}_3} = \frac{50}{50 + 149.9} = 0.25.$$

$$\text{s.s.c.} = \frac{1}{w} = \frac{1}{750} \frac{\text{kg}}{\text{kJ}} \left[\frac{\text{kJ}}{\text{s kW}}\right] \left[\frac{3600 \text{ s}}{\text{h}}\right] = \left[\frac{3600}{750}\right] = 4.8 \ \frac{\text{kg}}{\text{kW h}}.$$

Example 10.4

A steam power plant employing reheating of the steam after partial expansion is operated in conjunction with a district heating system under the following conditions:

Supply steam pressure 175 bar
Supply steam temperature 600 °C
Reheat steam pressure 20 bar
Reheat steam temperature 600 °C
Exhaust steam pressure 2 bar
Condensate temperature 80 °C

The energy rejected to cooling water is utilised in the district heating system. Cooling water leaves the condenser at 100 °C and is returned from the district at 35 °C. The isentropic efficiency (i.e. ratio of actual enthalpy drop to ideal enthalpy drop) of each expansion is 0.88.

Calculate

(a) the specific work transfer in each expansion,
(b) the specific energy supplied to the fluid in the boiler and in the reheating process,

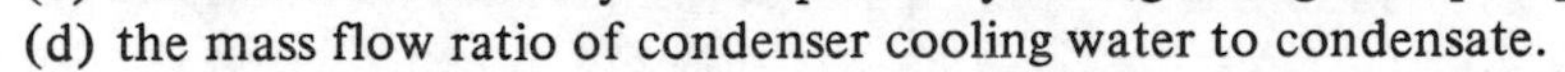
(c) the thermal efficiency of the power cycle neglecting feed pump work,
(d) the mass flow ratio of condenser cooling water to condensate.

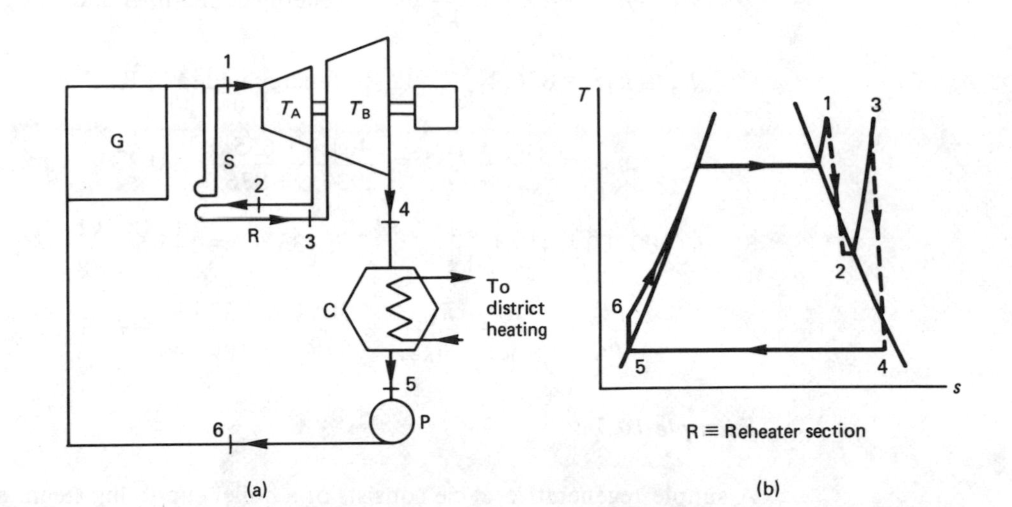

Answer

$$h_1 = \frac{3281 + 3268}{2} = 3274.5\ \frac{\text{kJ}}{\text{kg}}. \quad \text{(p. 8 of tables)}$$

$$s_1 = \frac{6.26 + 6.219}{2} = 6.2395 = s_{2,s} \quad \text{(i.e. wet vapour).} \quad \text{(p. 8)}$$
$$< s_{g,2}$$

$$x_{2,s} = \frac{s_{2,s} - s_{f,2}}{s_{fg,2}} = \frac{6.2395 - 2.447}{3.893} = 0.974. \quad \text{(p. 4)}$$

$$h_{2,s} = h_{f,2} + x_{2,s}h_{fg,2} = 909 + 0.974\ (1890) = 2750.2\ \frac{\text{kJ}}{\text{kg}}. \quad \text{(p. 4)}$$

$$h_1 - h_{2,s} = 524.3\ \frac{\text{kJ}}{\text{kg}}$$

$$h_1 - h_2 = \eta_T\,(h_1 - h_{2,s}) = 0.88\ (524.3) = 461.4\ \frac{\text{kJ}}{\text{kg}}$$

and $$h_2 = 3274.5 - 461.4 = 2813.1\ \frac{\text{kJ}}{\text{kg}}.$$

$$\left.\begin{aligned} h_3 &= 3690\ \frac{\text{kJ}}{\text{kg}}. \\ s_3 &= s_{4,s} = 7.701\ \frac{\text{kJ}}{\text{K kg}}. \end{aligned}\right\} \text{(p. 7)}$$

$$h_{4,s} = 2871 + \left(\frac{7.701 - 7.507}{7.708 - 7.507}\right)(2971 - 2871) = 2967.5\ \frac{\text{kJ}}{\text{kg}}. \quad \text{(p. 6)}$$

$$h_3 - h_{4,s} = 722.5\ \frac{\text{kJ}}{\text{kg}};$$

$$h_3 - h_4 = \eta_T\,(h_3 - h_{4,s}) = 0.88 \times 722.5\ \frac{\text{kJ}}{\text{kg}} = 635.8\ \frac{\text{kJ}}{\text{kg}}$$

and $$h_4 = 3690 - 635.8 = 3054.2\ \frac{\text{kJ}}{\text{kg}}.$$

$$h_5 = 334.9 \frac{kJ}{kg}. \quad \text{(p. 2)}$$

$$h_1 - h_5 = 2939.6 \frac{kJ}{kg} \qquad \text{(energy added in G and S);}$$

$$h_3 - h_2 = 876.9 \frac{kJ}{kg} \qquad \text{(energy added in R).}$$

$$\eta_{th} = \frac{{}_1w_2 + {}_3w_4}{{}_5q_1 + {}_2q_3} = \frac{461.4 + 635.8}{2939.6 + 876.9} = 0.287.$$

$$h_f(100\,°C) = 419.1 \frac{kJ}{kg}; \qquad h_f(35\,°C) = 146.55 \frac{kJ}{kg}. \quad \text{(p. 2)}$$

$$\frac{\dot{m}_w}{\dot{m}_s} = \frac{h_4 - h_5}{h_{f,100} - h_{f,35}} = \frac{3054.2 - 334.9}{419.1 - 146.55} = 9.98.$$

Example 10.5

A simple regenerative cycle consists of a boiler supplying steam at 40 bar, 500 °C, with exhaust to the condenser at 0.1 bar. The steam is reheated at 20 bar to 500 °C, and then is bled for regenerative feed heating at 10 bar.

The feed heater is of the closed type where the bled steam does not mix with the feed water but is throttled from 9 to 10 and then passed to the condenser as shown in the schematic.

The temperature of the feed water at 8 may be assumed to be 100 °C.

The isentropic efficiency of each section of the turbine expansion is 0.9, the feed pump work is small and state 9 may be assumed to be saturated liquid at 10 bar.

Draw a clear sketch of the cycle on the temperature–entropy field and calculate:

(a) the fraction of the mass flow which is bled for feed heating,
(b) the total power output for a mass flow rate at 1 of 3000 kg/min,
(c) the thermal efficiency of the cycle,
(d) the specific steam consumption.

Comment on the practicality of this cycle.

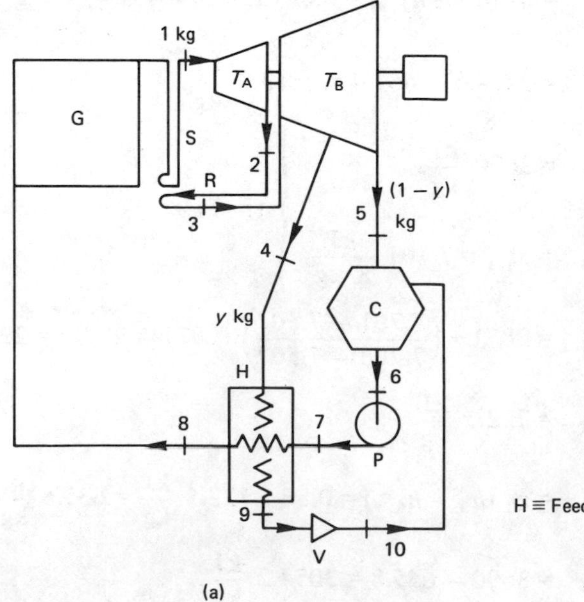

(a)

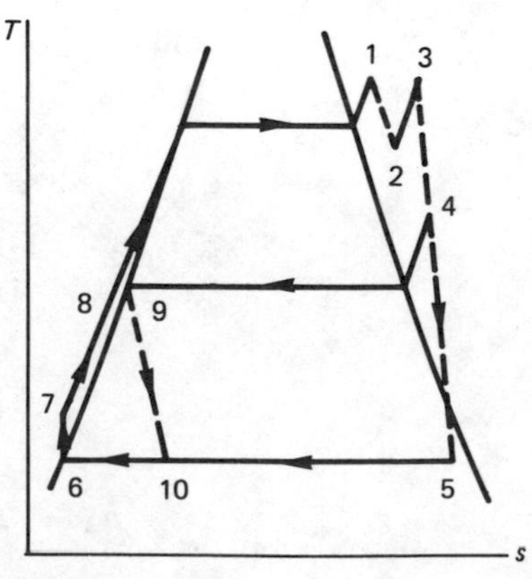

(b)

H ≡ Feed heater

Answer

Point	h	s
1	3445	7.089
2	3246	
3	3467	7.431
4	2955.9	6.948
5	2276.4	

Point	h
6	192
7	192
8	418.8
9	763.0
10	763.0

For determination of states see below

with h in $\frac{\text{kJ}}{\text{kg}}$ and s in $\frac{\text{kJ}}{\text{K kg}}$.

$h_1 = 3445; \quad s_1 = 7.089.$ (p. 7) $(= s_{2,s})$.

$$h_{2,s} = 3138 + \left(\frac{7.089 - 6.957}{7.126 - 6.957}\right)(3248 - 3138) = 3223.9. \quad \text{(p. 7 of tables)}$$

$h_1 - h_{2,s} = 221.1.$

$h_1 - h_2 = 0.9\,(221.1) = 199.0.$

$h_2 = 3246.$

$h_3 = 3467; \quad s_3 = 7.431.$ (p. 7)

$$h_{4,s} = 3158 + \left(\frac{7.431 - 7.301}{7.464 - 7.301}\right)(3264 - 3158) = 3242.5.$$

$h_3 - h_{4,s} = 224.5.$

$h_3 - h_4 = 0.9\,(224.5) = 202.1.$

$h_4 = 2955.9.$

$$s_4 = 6.926 + \left(\frac{2955.9 - 2944}{3052 - 2944}\right)(7.124 - 6.926) = 6.948.$$

$$x_{5,s} = \frac{6.948 - 0.649}{7.5} = 0.84. \quad \text{(p. 3)}$$

$h_{5,s} = 192 + 0.84\,(2392) = 2201.3.$ (p. 3)

$h_4 - h_{5,s} = 755.$

$h_4 - h_5 = 0.9\,(755) = 679.5.$

$h_5 = 2276.4.$

$$x_5 = \frac{2276.4 - 192}{2392} = 0.871. \quad \text{(p. 3)}$$ (Too low a value since this will lead to blade erosion)

$h_6 = h_{f,5} = 192$ (p. 3) $= h_7$ (neglect feed pump work).

$h_8 = h_{f,100} = 419.1$ (p. 2)

$h_9 = h_{10} = 763$ (p. 4) $= h_f$ at 10 bar. (No change in h in throttling.)

Heater H – *energy balance*

$$y\,(h_4 - h_9) = 1\,(h_8 - h_6)$$

$$y = \frac{419.1 - 193}{2955.9 - 763} = \frac{226.1}{2192.9} = 0.103.$$

$$\dot{W}_{\text{turbine}} = \dot{m}_{T,A}(h_1 - h_2) + \dot{m}_{T,B}(h_3 - h_4) + \dot{m}_{T,B}(h_4 - h_5)$$

$$= 3000 \frac{\text{kg}}{\text{mm}} \left[\frac{\text{min}}{60 \text{ s}}\right] \left\{(199) + (202.1) + (1 - 0.103)(679.5)\right\} \frac{\text{kJ}}{\text{kg}} \left[\frac{\text{MW}}{10^3 \text{ kW}}\right]$$

$$= 50.53 \text{ MW}.$$

$${}_8q_1 + {}_2q_3 = (h_1 - h_8) + (h_3 - h_2) = (3445 - 419.1) + (3467 - 3246)$$

$$= 3246.9 \frac{\text{kJ}}{\text{kg}}.$$

$$\dot{Q}_{\text{added}} = m\,({}_8q_1 + {}_2q_3) = 300 \frac{\text{kg}}{\text{min}} \left[\frac{\text{min}}{60 \text{ s}}\right] (3246.9) \frac{\text{kJ}}{\text{kg}} \left[\frac{\text{MW}}{10^3 \text{ kW}}\right]$$

$$\dot{Q}_{\text{added}} = 162.35 \text{ MW}.$$

$$\eta_{\text{thermal}} = \frac{50.53}{162.35} = 0.311.$$

$$\text{Specific steam consumption} = \frac{1}{w}$$

$$= \frac{1}{199 \frac{\text{kJ}}{\text{kg}} + 202.1 \frac{\text{kJ}}{\text{kg}} + \left(\frac{679.5}{1 - 0.103}\right) \frac{\text{kJ}}{\text{kg}}} \left[\frac{\text{kJ}}{\text{kW s}}\right] \left[\frac{3600 \text{ s}}{\text{h}}\right]$$

$$= 3.107 \frac{\text{kg}}{\text{kW h}}.$$

Example 10.6

A steam power plant operates on the Rankine cycle with reheat. The supply steam expands in a high-pressure turbine from 50 bar and 450 °C to 2 bar. The steam is returned to the boiler and reheated to a temperature of 350 °C (at 2 bar) and expanded in a low-pressure turbine to a condenser pressure of 0.055 bar. The isentropic efficiency of each turbine is 0.9. The condensate is undercooled to 26 °C and the isentropic efficiency of the feed pump compression is 0.7.

Draw a T–s diagram for the plant.

Determine the condition of the steam at exhaust from each turbine and obtain the cycle efficiency.

What is the benefit of reheating the steam?

Answer

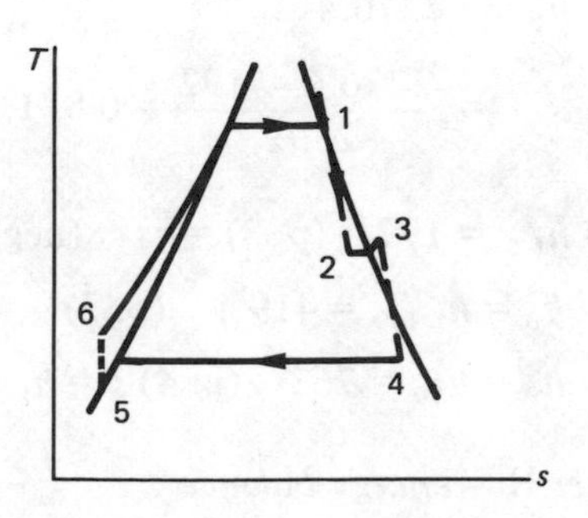

Values of h in $\frac{\text{kJ}}{\text{kg}}$ and s in $\frac{\text{kJ}}{\text{K kg}}$.

Tables, page 7: $h_1 = 3316;$
$s_1 = 6.818 = s_{2,s}.$

page 4:

$$\begin{cases} x_{2,s} = \dfrac{s_{2,s} - s_{f,2}}{s_{fg}} = \dfrac{6.818 - 1.53}{5.597} = 0.945; \\ h_{2,s} = h_{f,2} + x_2 (h_{fg,2}) = 505 + 0.945\,(2202) \\ h_{2,s} = 258.4. \end{cases}$$

Thus $h_1 - h_{2,s} = 730.6$

and $h_1 - h_2 = {}_1w_2 = 0.9 \times 730.6 = 657.5.$

Also $h_2 = 3316 - 657.5 = 2658.5$

and $x_2 = \dfrac{2658.5 - 505}{2202} = 0.978.$

page 6:

$$\begin{cases} h_3 = \dfrac{3072 + 3277}{2} = 3174.5 \\ s_3 = \dfrac{7.892 + 8.221}{2} = 8.009 = s_{4,s}. \end{cases}$$

page 3:

$$\begin{cases} x_{4,s} = \dfrac{s_{4,s} - s_{f,4}}{s_{fg,4}} = \dfrac{8.009 - 0.5}{7.86} = 0.955 \\ h_{4,s} = 145 + 0.955\,(2419) = 2456. \end{cases}$$

Thus $h_3 - h_{4,s} = 718.5$

and $h_3 - h_4 = 0.9 \times 718.5 = 646.7 = {}_3w_4.$

$$h_4 = 3174.5 - 646.7 = 2527.8.$$

$$x_4 = \frac{2527.8 - 145}{2419} = 0.985.$$

page 2 at 26 °C, $h_5 = h_f = 108.9.$

$$h_{6,s} - h_5 = \int_{p,5}^{p,6} v\,dp \simeq v_f\,.\,\Delta p = \underset{\uparrow \text{ p. 10}}{0.001\,033}\ \frac{m^3}{kg}\,(5000 - 5.5)\ \frac{kN}{m^2} = 5.16.$$

$$h_6 - h_5 = \frac{5.16}{0.7} = 7.37 \quad \left(\text{or } {}_5w_6 = -7.37\ \frac{kJ}{kg}\right);$$

$$h_6 = 108.9 + 7.37 = 116.27.$$

$$w_{net} = {}_1w_2 + {}_3w_9 + {}_5w_6 = (h_1 - h_2) + (h_3 - h_4) + (h_5 - h_6)$$
$$= 1296.83.$$

$$q_{+ve} = {}_6q_1 + {}_2q_3 = (h_1 - h_6) + (h_3 - h_2)$$
$$= (3316 - 116.27) + (3174.5 - 2658.4) = 3715.83.$$

$$\eta_{th} = \frac{w_{net}}{q_{+ve}} = \frac{1296.83}{3715.83} = 0.349.$$

Reheating results in a drier exhaust steam condition when leaving the turbine and obviates blade erosion problems.

Exercises

1 A vapour power cycle consists of a generator delivering steam at 50 bar, 350 °C, a turbine of isentropic efficiency 0.85 with an exhaust pressure of 1 bar, a condenser and a feed pump.

Assuming saturated liquid at entry to the boiler, calculate the Rankine efficiency of the cycle (both ignoring the feed pump work and then taking this into account), the specific steam consumption and the specific heat transfer in the condenser.

(0.235, 0.233, 5.773 kg/(kW h), 2029.4 kJ/kg)

2 A vapour power cycle has turbine inlet conditions of 40 bar, 500 °C and the turbine exhausts at 0.5 bar. The steam is reheated at 20 bar to 500 °C and each section of the turbine has an isentropic efficiency of 0.87.

Calculate (ignoring the feed pump work) the Rankine efficiency assuming saturated liquid at entry to the steam generator. (0.278)

3 Calculate the Rankine efficiency for a vapour power cycle with turbine inlet conditions of 60 bar, 350 °C, if the turbine expands with isentropic efficiency of 0.84 to an exhaust pressure 0.1 bar.

Compare this with the efficiency of a Carnot cycle between the same overall thermodynamic temperature limits, assuming that the liquid is saturated at entry to the steam generator in the Rankine cycle.

Calculate also the specific steam consumption for each cycle, making due allowance for the compression work done in the Carnot cycle.

What are the objections to each of these cycles in practice?

$$\left(0.306,\ 0.488,\ \text{s.s.c.}_R = 4.124\ \frac{\text{kg}}{\text{kW h}},\ \text{s.s.c.}_C = 8.244\ \frac{\text{kg}}{\text{kW h}}\right)$$

11 Gas Power Cycles

11.1 Introduction

This chapter is exclusively devoted to ideal air standard cycles which assume air as the working fluid for which the perfect gas laws may be assumed to hold true.

These cycles serve as suitable ideal standards with which to compare real engine performance.

There are four criteria of significance in these cycles, as follows:

(a) *Thermal efficiency*, η_{th}:

$$\eta_{th} = \frac{\text{Net work output}}{\text{Positive heat in cycle}}.$$

Note that the denominator implies that heat may be added in more than one process in the cycle which for example occurs in the dual combustion cycle (heat is added partly at constant volume and partly at constant pressure).

(b) *Work ratio*, r_w:

$$r_w = \frac{\text{Net work output}}{\text{Positive work in cycle}},$$

a criterion showing a cycle's susceptibility to irreversibility. (Note that work ratio is nearly unity with the Rankine cycle in the previous chapter but different values result in this one.)

(c) *Mean effective pressure*, m.e.p.:

$$\text{m.e.p.} = \frac{\text{Net work output}}{\text{Swept Volume}}.$$

This is a power criterion and mean effective pressure is that constant pressure assumed to act over the entire swept volume of the cycle that will yield the same power output as the actual cycle.

(d) *Compression ratio*, r:

$$r = \frac{\text{Swept volume} + \text{Clearance volume}}{\text{Clearance volume}}$$

(circa 8 to 10 for spark-ignition engines, 16 to 20 for compression engines).

11.2 Worked Examples

Example 11.1

In a Carnot cycle 1 kg of air is compressed isothermally from 1 bar to 4 bar, the temperature being 15 °C. The maximum temperature reached in the cycle is 400 °C.

Find the thermal efficiency, mean effective pressure and work ratio.

The Carnot cycle consists of two isothermals and two isentropes as shown in

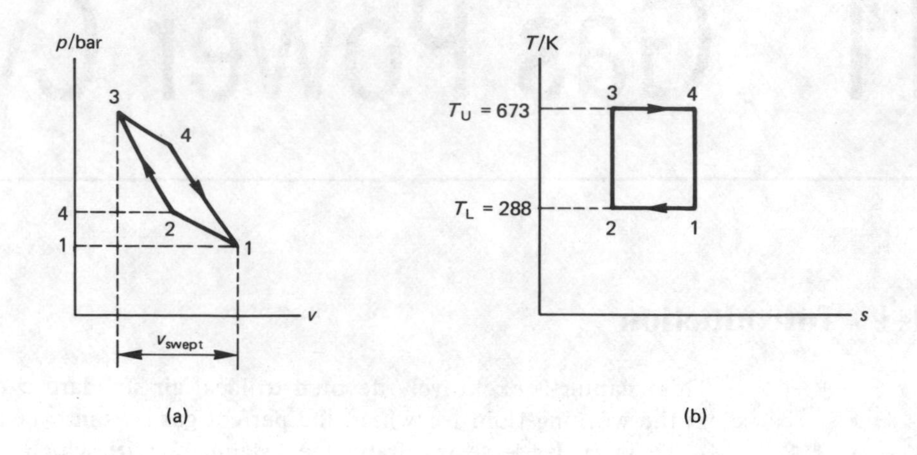

the diagrams and has the highest thermal efficiency between the two extreme limits of temperature.

Answer

Thermal efficiency $\eta_{th} = \dfrac{w_{net}}{q_{+ve}} = \dfrac{\Sigma w}{q_{+ve}} = \dfrac{\Sigma q}{q_{+ve}}$ $(\Sigma w = \Sigma q)$

$$= \frac{{}_3q_4 + {}_1q_2}{{}_3q_4} = \frac{T_U (s_4 - s_3) + T_L (s_2 - s_1)}{T_U (s_4 - s_3)}$$

$$= \frac{T_U - T_L}{T_U} \quad \text{since } (s_2 - s_1) = -(s_4 - s_3)$$

$$= \frac{673 - 288}{673} = 0.572.$$

$$v_1 = \frac{p_2 v_2 T_1}{T_2 p_1} = 4v_2.$$

$$v_3 = v_2 \left(\frac{T_2}{T_3}\right)^{1/(\gamma-1)} = \frac{v_1}{4}\left(\frac{288}{673}\right)^{1/0.4} = 0.0299\ v_1 \qquad (\gamma = 1.4 \text{ for air; p. 24})$$

$$v_1 - v_3 = v_{swept} = (1 - 0.0299)\ v_1 = 0.9701 \times \frac{RT_1}{p_1}$$

$$= \frac{0.9701 \times 0.287\ \dfrac{kJ}{kg\ K} \times 288\ K}{100\ \dfrac{kN}{m^2}} = 0.802\ \frac{m^3}{kg}.$$

$$s_2 - s_1 = c_p \ln \frac{T_2}{T_1} - R \ln \frac{p_2}{p_1} = -R \ln \frac{p_2}{p_1} \qquad (T_2 = T_1 = T_L)$$

$$= -0.287\ \frac{kJ}{kg\ K} \ln 4 = -0.3979\ \frac{kJ}{K\ kg}.$$

$$s_4 - s_3 = +0.3979\ \frac{kJ}{K\ kg}.$$

$$q_{+ve} = T_U\ (s_4 - s_3) = 673\ K\ (+0.3979)\ \frac{kJ}{K\ kg} = 267.8\ \frac{kJ}{kg}.$$

$$\text{Mean effective pressure} = \frac{w_{net}}{v_1 - v_3} = \frac{\eta_{th}\, q_{+ve}}{v_1 - v_3}$$

$$= \frac{0.572 \times 267.8 \ \dfrac{\text{kJ}}{\text{kg}}}{0.802 \ \dfrac{\text{m}^3}{\text{kg}}} \left[\frac{\text{kN m}}{\text{kJ}}\right] = 191.0 \ \frac{\text{kN}}{\text{m}^2} = 1.91 \text{ bar.}$$

$${}_4w_1 = e_4 - e_1 = c_v(T_4 - T_1) = 0.718 \ \frac{\text{kJ}}{\text{kg K}} (673 - 288) \text{ K} = 276.43 \ \frac{\text{kJ}}{\text{kg}}$$

$$p_3 = p_2 \left(\frac{T_3}{T_2}\right)^{\gamma/(\gamma-1)} = 4 \text{ bar} \left(\frac{673}{288}\right)^{3.5} = 78.02 \text{ bar.}$$

$$p_4 = \frac{p_3 v_3}{v_4} = \frac{p_3 v_3}{v_1 \left(\dfrac{T_1}{T_4}\right)^{1/(\gamma-1)}} = \frac{78.02 \text{ bar} \times 0.0299}{\left(\dfrac{288}{673}\right)^{2.5}} = 19.47 \text{ bar.}$$

$${}_3w_4 = {}_3q_4 = T_U(s_4 - s_3) = -T_U R \ln \frac{p_4}{p_3} = T_U R \ln \frac{p_3}{p_4}$$

$$= 673 \text{ K} \times 0.287 \ \frac{\text{kJ}}{\text{kg K}} \ln \frac{78.02}{19.47} = 268.09 \ \frac{\text{kJ}}{\text{kg}}.$$

$$w_{+ve} = {}_3w_4 + {}_4w_1 = 276.43 + 268.09 = 544.52 \ \frac{\text{kJ}}{\text{kg}}.$$

$$r_w = \frac{w_{net}}{w_{+ve}} = \frac{153.2}{544.52} = 0.281.$$

This value of work ratio is low once again showing the susceptibility of this cycle to irreversibilities in practice.

Example 11.2

In an air standard Otto cycle the fluid is at 1 bar and 288 K at the start of compression. The volume compression ratio is 8 and the maximum cycle temperature is 2500 K.

Calculate the thermal efficiency, mean effective pressure and work ratio.

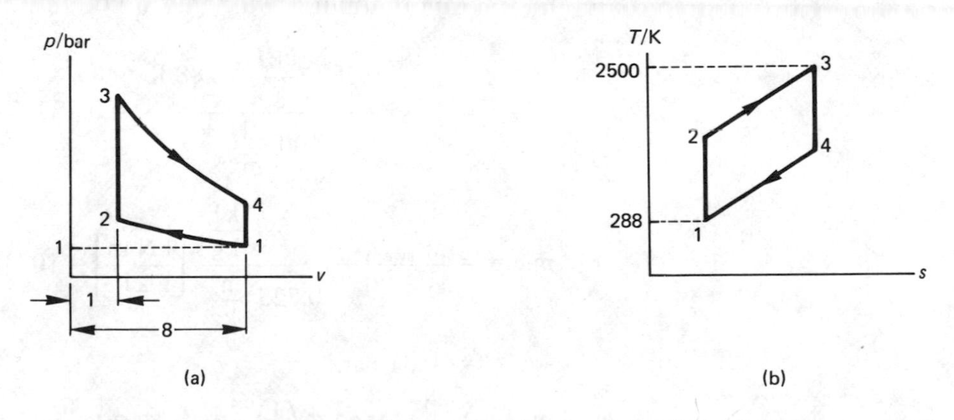

As shown in the diagrams, the Otto cycle consists of the following operations in order:

(a) isentropic compression, $1 \rightarrow 2$;
(b) isochoric (constant-volume) heating, $2 \rightarrow 3$;
(c) isentropic expansion, $3 \rightarrow 4$;
(d) isochoric cooling, $4 \rightarrow 1$.

Answer

$$T_2 = T_1 \left(\frac{v_1}{v_2}\right)^{\gamma-1} = T_1 \times r^{\gamma-1} \qquad \left(r = \frac{v_1}{v_2} = \text{compression ratio}\right).$$

$$T_4 = T_3 \left(\frac{v_3}{v_4}\right)^{\gamma-1} = T_3 \times r^{1-\gamma} \qquad \left(\frac{v_3}{v_4} = \frac{v_2}{v_2} = \frac{1}{r}\right).$$

$$\eta_{th} = \frac{w_{net}}{q_{+ve}} = \frac{{}_3w_4 + {}_1w_2}{{}_2q_3}$$

since ${}_2w_3 = {}_4w_1 = 0$
and ${}_3q_4 = {}_1q_2 = 0$
and ${}_4q_1$ is –ve.

$$\therefore \eta_{th} = \frac{\frac{R}{\gamma-1}\left\{(T_3 - T_4) + (T_1 - T_2)\right\}}{c_v(T_3 - T_2)} = \frac{(T_3 - T_2) + (T_1 - T_4)}{(T_3 - T_2)}$$

$$= 1 + \frac{T_1 - T_4}{T_3 - T_2}.$$

$$T_2 = T_1 \left(\frac{v_1}{v_2}\right)^{\gamma-1} = T_1 \times r^{\gamma-1}; \qquad T_4 = T_3 \left(\frac{v_3}{v_4}\right)^{\gamma-1} = T_2 \times r^{1-\gamma};$$

$$\therefore \eta_{th} = 1 + \frac{(T_1 - T_3 r^{1-\gamma})}{(T_3 - T_1 r^{\gamma-1})} = 1 - r^{1-\gamma}\left\{\frac{T_3 - T_1 r^{\gamma-1}}{T_3 - T_1 r^{\gamma-1}}\right\}$$

$$= 1 - \frac{1}{r^{\gamma-1}} = 1 - \frac{1}{8^{0.4}} = 0.565.$$

$$T_2 = T_1 r^{\gamma-1} = 288 \text{ K } (8)^{0.4} = 661.7 \text{ K}.$$

$${}_2q_3 = c_v(T_3 - T_2) = 0.718 \frac{\text{kJ}}{\text{kg K}} (2500 - 661.7) \text{ K} = 1319.9 \frac{\text{kJ}}{\text{kg}}.$$

$$w_{net} = \eta_{th} \times {}_2q_3 = 0.565 \times 1319.9 = 745.7 \frac{\text{kJ}}{\text{kg}}.$$

$$v_1 - v_2 = v_{swept} = 8v_2 - v_2 = 7v_2 = \frac{7}{8} v_1 = \frac{7}{8}\frac{RT_1}{p_1}$$

$$= \frac{7}{8} \times \frac{0.287 \frac{\text{kJ}}{\text{kg K}} \times 288 \text{ K}}{100 \frac{\text{kN}}{\text{m}^2}} = 0.723 \frac{\text{m}^3}{\text{kg}}.$$

$$\text{m.e.p.} = \frac{w_{swept}}{v_1 - v_2} = \frac{745.7 \frac{\text{kJ}}{\text{kg}}}{0.723 \frac{\text{m}^3}{\text{kg}}} \left[\frac{\text{kN m}}{\text{kJ}}\right] = 1031.4 \frac{\text{kN}}{\text{m}^2}$$

$$= 10.31 \text{ bar}.$$

$$T_4 = T_3 \left(\frac{v_3}{v_4}\right)^{\gamma-1} = 2500 \text{ K} \left(\frac{1}{8}\right)^{0.4} = 1088.2 \text{ K}.$$

$${}_3w_4 = w_{+ve} = \frac{R(T_3 - T_4)}{\gamma - 1} = \frac{0.287}{0.4} \frac{\text{kJ}}{\text{kg K}} (2500 - 1088.2) \text{ K}$$

$$w_{+ve} = 1013 \frac{\text{kJ}}{\text{kg}}.$$

$$r_w = \text{work ratio} = \frac{w_{\text{net}}}{w_{+ve}} = \frac{745.7 \text{ kJ/kg}}{1013 \text{ kJ/kg}} = 0.736.$$

Note that the Carnot efficiency between the same temperature limits of 2500 K and 288 K is 0.885 and the Otto cycle efficiency is 0.565 but the value of work ratio is much higher than in the Carnot cycle which means that irreversibilities will have a correspondingly smaller effect on this cycle in practice.

Example 11.3

An air standard diesel cycle consists of the following operations in order:

(a) isentropic compression from 1 bar, 288 K, to a maximum pressure of 90 bar;
(b) heat addition at constant pressure till the specific volume is doubled;
(c) isentropic expansion to the initial volume;
(d) isochoric cooling to the initial state.

Calculate the volume compression ratio, thermal efficiency, mean effective pressure and work ratio.

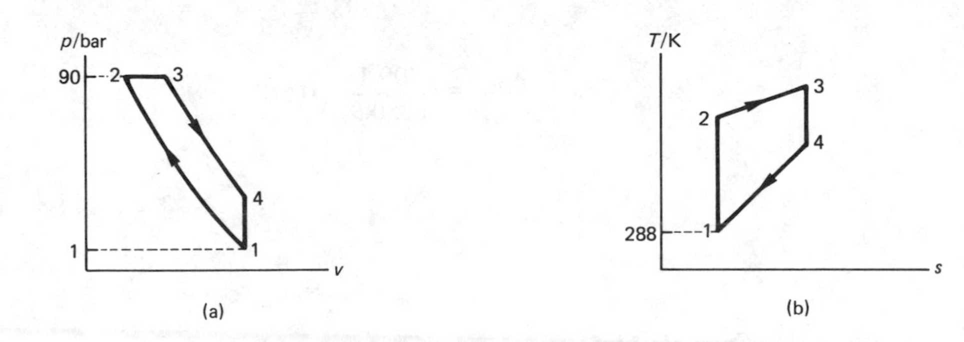

Answer

$$T_2 = T_1 \left(\frac{p_2}{p_1}\right)^{(\gamma-1)/\gamma} = 288 \text{ K } (90)^{0.286} = 1043.1 \text{ K}.$$

$$\frac{v_1}{v_2} = \left(\frac{T_2}{T_1}\right)^{1/(\gamma-1)} = \left(\frac{1043.1}{288}\right)^{2.5} = 24.96 = r.$$

$$\frac{T_3}{T_2} = \frac{v_3}{v_2} = 2.$$

$$T_3 = 2 \times 1043.1 = 2086.2 \text{ K}.$$

$${}_2q_3 = c_p (T_3 - T_2) = 1.005 \frac{\text{kJ}}{\text{kg K}} (1043.1) \text{ K} = 1048.3 \frac{\text{kJ}}{\text{kg}}.$$

$${}_2w_3 = R (T_3 - T_2) = 0.287 \frac{\text{kJ}}{\text{kg K}} (1043.1) \text{ K} = 299.4 \frac{\text{kJ}}{\text{kg}}.$$

$$T_4 = T_3 \left(\frac{v_3}{v_4}\right)^{\gamma-1} = 2086.2 \text{ K} \left[\frac{2}{24.96}\right]^{0.4} = 760.1 \text{ K}.$$

$${}_3w_4 = \frac{R (T_3 - T_4)}{\gamma - 1} = \frac{0.287 \frac{\text{kJ}}{\text{kg K}} (2086.2 - 760.1) \text{ K}}{0.4} = 951.5 \frac{\text{kJ}}{\text{kg}}.$$

$$ {}_1w_2 = \frac{R(T_1 - T_2)}{\gamma - 1} = \frac{0.287 \, \dfrac{\text{kJ}}{\text{kg K}} \, (288 - 1043.1) \, \text{K}}{0.4} = -541.8 \, \frac{\text{kJ}}{\text{kg}} . $$

$$ w_{+\text{ve}} = {}_2w_3 + {}_3w_4 = 1250.9 \, \frac{\text{kJ}}{\text{kg}} . $$

$$ w_{\text{net}} = {}_2w_3 + {}_3w_4 + {}_1w_2 = 709.1 \, \frac{\text{kJ}}{\text{kg}} \qquad ({}_4w_1 = 0). $$

$$ \eta_{\text{th}} = \frac{w_{\text{net}}}{q_{+\text{ve}}} = \frac{709.1}{1048.3} = 0.676. $$

$$ v_1 - v_2 = v_{\text{swept}} = 23.96 v_2 = \frac{23.96}{24.96} v_1 = \frac{23.96}{24.96} \frac{RT_1}{p_1} $$

$$ = \frac{23.96}{24.96} \times 0.287 \, \frac{\text{kJ}}{\text{kg K}} \times \frac{288 \, \text{K}}{100 \, \dfrac{\text{kN}}{\text{m}^2}} = 0.793 \, \frac{\text{m}^3}{\text{kg}} . $$

$$ \text{m.e.p.} = \frac{w_{\text{net}}}{v_1 - v_2} = \frac{709.1 \, \dfrac{\text{kJ}}{\text{kg}}}{0.793 \, \dfrac{\text{m}^3}{\text{kg}}} \left[\frac{\text{kN m}}{\text{kJ}} \right] = 894.2 \, \frac{\text{kN}}{\text{m}^2} = 8.942 \text{ bar}. $$

$$ r_{\text{w}} = \frac{w_{\text{net}}}{w_{+\text{ve}}} = \frac{709.1}{1250.9} = 0.567. $$

Example 11.4

A dual-combustion cycle consists of the following operations in order:

(a) isentropic compression from 1 bar, 288 K, through a volume compression ratio of 20;
(b) heat addition of 1500 kJ/kg in two parts, two thirds being at constant volume, followed by one third at constant pressure;
(c) isentropic expansion to the original volume;
(d) isochoric cooling to the initial state.

Calculate the thermal efficiency, mean effective pressure and work ratio for the cycle.

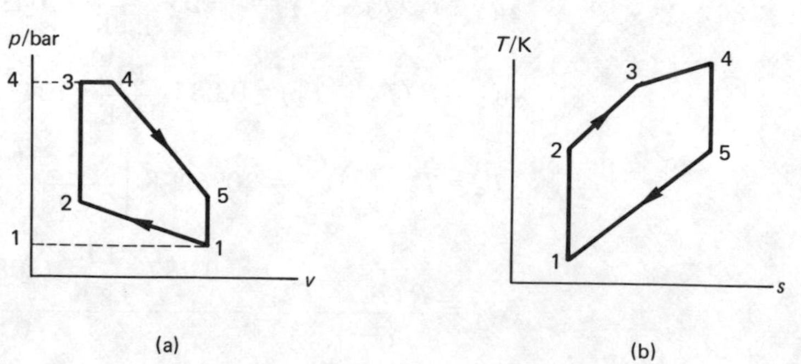

Answer

$$T_2 = T_1 \left(\frac{v_1}{v_2}\right)^{\gamma-1} = 288 \text{ K } (20)^{0.4} = 954.6 \text{ K}.$$

$$T_3 - T_2 = \frac{{}_2q_3}{c_v} = \frac{1000 \dfrac{\text{kJ}}{\text{kg}}}{0.718 \dfrac{\text{kJ}}{\text{kg K}}} = 1392.8 \text{ K}.$$

$$T_3 = 2347.4 \text{ K}.$$

$$T_4 - T_3 = \frac{{}_3q_4}{c_p} = \frac{500 \dfrac{\text{kJ}}{\text{kg}}}{1.005 \dfrac{\text{kJ}}{\text{kg K}}} = 497.5 \text{ K}.$$

$$T_4 = 2844.9 \text{ K}.$$

$$v_4 = v_3 \left(\frac{T_4}{T_3}\right) = \frac{v_1}{20} \frac{2844.9}{2347.4} = 0.0606 v_1 = 0.0606 v_5 .$$

$$T_5 = T_4 \left(\frac{v_4}{v_5}\right)^{\gamma-1} = 2844.9 \text{ K } (0.0606)^{0.4} = 926.9 \text{ K}.$$

$${}_3w_4 = R\,(T_4 - T_3) = 0.287 \frac{\text{kJ}}{\text{kg K}} (2844.9 - 2347.4) \text{ K} = 142.8 \frac{\text{kJ}}{\text{kg}} .$$

$${}_4w_5 = \frac{R\,(T_4 - T_5)}{\gamma - 1} = \frac{0.287 \dfrac{\text{kJ}}{\text{kg K}} (2844.9 - 926.9) \text{ K}}{0.4} = 1376.2 \frac{\text{kJ}}{\text{kg}} .$$

$$w_{+\text{ve}} = 142.8 + 1376.2 = 1519 \frac{\text{kJ}}{\text{kg}} .$$

$${}_1w_2 = \frac{R\,(T_1 - T_2)}{\gamma} = \frac{0.287 \dfrac{\text{kJ}}{\text{kg K}} (288 - 954.6) \text{ K}}{0.4} = -\,478.3 \frac{\text{kJ}}{\text{kg}} .$$

$$w_{\text{net}} = 142.8 + 1376.2 - 478.3 = 1040.7 \frac{\text{kJ}}{\text{kg}} .$$

$$\eta_{\text{th}} = \frac{w_{\text{net}}}{q_{+\text{ve}}} = \frac{1040.7}{1500} = 0.694.$$

$$v_1 - v_2 = 19 v_2 = \frac{19}{20} v_1 = \frac{19}{20} \frac{RT_1}{p_1} = \frac{19}{20} \times \frac{0.287 \dfrac{\text{kJ}}{\text{kg K}} \times 288 \text{ K}}{100 \dfrac{\text{kN}}{\text{m}^2}} ;$$

$$v_1 - v_2 = v_{\text{swept}} = 0.785 \frac{\text{m}^3}{\text{kg}} .$$

$$\text{m.e.p.} = \frac{w_{\text{net}}}{v_1 - v_2} = \frac{1040.7 \dfrac{\text{kJ}}{\text{kg}}}{0.785 \dfrac{\text{m}^3}{\text{kg}}} \left[\frac{\text{kN m}}{\text{kJ}}\right] = 1325 \frac{\text{kN}}{\text{m}^2}$$

m.e.p. = 13.25 bar.

$$r_w = \frac{w_{\text{net}}}{w_{+\text{ve}}} = \frac{1040.7}{1519} = 0.685.$$

Example 11.5

An air standard Joule cycle consists of the following operations in order:

(a) isentropic compression from 1 bar, 288 K, through a pressure ratio of 4;
(b) isobaric heat addition of 1000 kJ/kg;
(c) isentropic expansion to the initial pressure;
(d) isobaric cooling to the initial state.

Calculate the thermal efficiency, mean effective pressure and work ratio for the cycle.

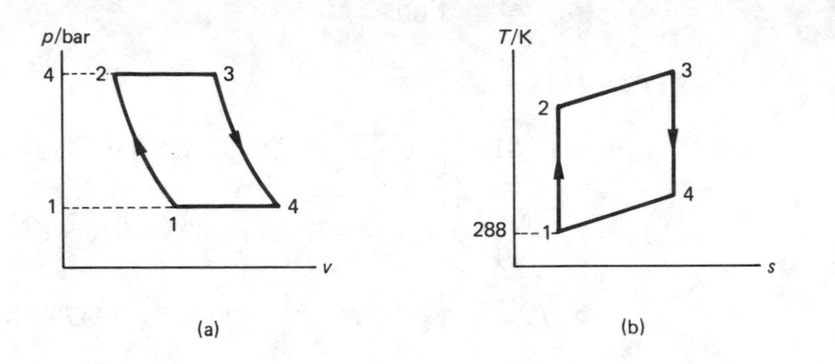

Answer

$$T_2 = T_1 \left(\frac{p_2}{p_1}\right)^{(\gamma-1)/\gamma} = 288 \text{ K } (4)^{0.286} = 428.1 \text{ K}.$$

$$T_3 = T_2 + \left(\frac{{}_2q_3}{c_p}\right) = 428.1 \text{ K} + \frac{1000 \dfrac{\text{kJ}}{\text{kg}}}{1.005 \dfrac{\text{kJ}}{\text{kg K}}} = 1423.1 \text{ K}.$$

$$T_4 = T_3 \left(\frac{p_4}{p_3}\right)^{(\gamma-1)/\gamma} = 1423.1 \text{ K } (\tfrac{1}{4})^{0.286} = 957.3 \text{ K}.$$

$${}_4q_1 = c_p\,(T_1 - T_4) = 1.005 \frac{\text{kJ}}{\text{kg K}} (288 - 957.3) \text{ K} = -\,672.6 \frac{\text{kJ}}{\text{kg}}.$$

$$w_{\text{net}} = \Sigma w = \Sigma q = {}_2q_3 + {}_4q_1 = 1000 - 672.6 = 327.4 \frac{\text{kJ}}{\text{kg}} \qquad ({}_1q_2 = {}_3q_4 = 0).$$

$$\eta_{\text{th}} = \frac{w_{\text{net}}}{q_{+\text{ve}}} = \frac{327.4}{1000} = 0.327.$$

$$v_4 = v_1 \left(\frac{T_4}{T_1}\right) = v_1 \left(\frac{957.3}{288}\right) = 3.324 v_1 = 3.324 \frac{RT_1}{p_1}$$

$$= 3.324 \times \frac{0.287 \dfrac{\text{kJ}}{\text{kg K}} \times 288 \text{ K}}{100 \dfrac{\text{kN}}{\text{m}^2}} = 2.747 \frac{\text{m}^3}{\text{kg}}.$$

$$v_2 = v_1 \left(\frac{p_1}{p_2}\right)^{1/\gamma} = \frac{RT_1}{p_1}\left(\frac{p_1}{p_2}\right)^{1/\gamma} = \frac{0.287 \dfrac{\text{kJ}}{\text{kg K}} \times 288 \text{ K}}{100 \dfrac{\text{kN}}{\text{m}^2}} (\tfrac{1}{4})^{1/1.4} = 0.307 \frac{\text{m}^3}{\text{kg}}.$$

$$v_4 - v_2 = v_{\text{swept}} = 2.44\ \frac{\text{m}^3}{\text{kg}}.$$

$$\text{m.e.p.} = \frac{w_{\text{net}}}{v_{\text{swept}}} = \frac{327.4\ \frac{\text{kJ}}{\text{kg}}}{2.44\ \frac{\text{m}^3}{\text{kg}}} = 134\ \frac{\text{kN}}{\text{m}^2} = 1.34\ \text{bar}.$$

$$w_{\text{net}} = {}_2w_3 + {}_3w_4 = R\,(T_3 - T_2) + \frac{R}{\gamma - 1}\,(T_3 - T_4)$$

$$= 0.287\ \frac{\text{kJ}}{\text{kg K}}\ (1423.1 - 428.1)\ \text{K} + \frac{0.287\ \frac{\text{kJ}}{\text{kg K}}}{0.4}\ (1423.1 - 957.3)\ \text{K}$$

$$= 619.8\ \frac{\text{kJ}}{\text{kg}}.$$

$$r_{\text{w}} = \frac{w_{\text{net}}}{w_{+\text{ve}}} = \frac{327.4}{619.8} = 0.528.$$

Example 11.6

The following data refer to an air standard dual-combustion cycle:

Minimum pressure 1 bar
Maximum pressure 60 bar
Minimum temperature 15 °C
Compression ratio 15 : 1
Thermal efficiency 58 per cent
Mean effective pressure 10 bar

Calculate the maximum cycle temperature and the work ratio.

Answer

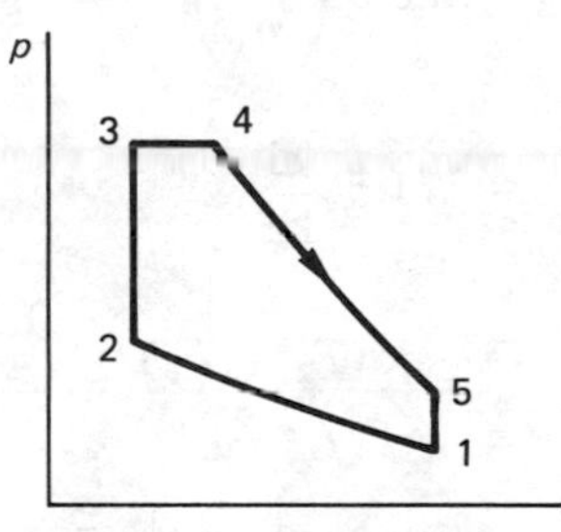

$$T_2 = T_1 \left(\frac{v_1}{v_2}\right)^{\gamma - 1} = 288\ \text{K}\ (15)^{0.4} = 850.8\ \text{K}.$$

$$p_2 = p_1 \left(\frac{v_1}{v_2}\right)^{\gamma} = 1\ \text{bar}\ (15)^{1.4} = 44.3\ \text{bar}.$$

$$T_3 = T_2 \left(\frac{p_3}{p_2}\right) = 850.8\ \text{K} \times \frac{60}{44.3} = 1152\ \text{K}.$$

$$\eta_{\text{th}} = \frac{w_{\text{net}}}{q_{+\text{ve}}} = \frac{R\,(T_4 - T_3) + \frac{R}{\gamma - 1}\,(T_4 - T_5) - \frac{R}{\gamma - 1}\,(T_2 - T_1)}{c_v\,(T_3 - T_2) + c_p\,(T_4 - T_3)}$$

(T_4, T_5 unknown).

$$\text{m.e.p.} = \frac{w_{\text{net}}}{v_1 - v_2} = \frac{R(T_4 - T_3) + \dfrac{R}{\gamma - 1}(T_4 - T_5) - \dfrac{R}{\gamma - 1}(T_2 - T_1)}{(v_1 - v_2)}$$

$$= \frac{R(T_4 - T_3) + \dfrac{R}{\gamma - 1}(T_4 - T_5) - \dfrac{R}{\gamma - 1}(T_2 - T_1)}{v_1\left(1 - \dfrac{v_2}{v_1}\right)}$$

$$= \frac{R(T_4 - T_3) + \dfrac{R}{\gamma - 1}(T_4 - T_5) - \dfrac{R}{\gamma - 1}(T_2 - T_1)}{\dfrac{RT_1}{p_1}\left(1 - \dfrac{1}{15}\right)}.$$

The numerators of both the η_{th} and m.e.p. expressions are identical.

Thus $$\eta_{\text{th}}\left[c_v(T_3 - T_2) + c_p(T_4 - T_3)\right] = \text{m.e.p.}\,\frac{RT_1}{p_1}\left(\frac{14}{15}\right)$$

or $$0.58\left[0.718\,\frac{\text{kJ}}{\text{kg K}}\,(1152 - 850.8)\text{ K} + 1.005\,\frac{\text{kJ}}{\text{kg K}}\,(T_4 - 1152)\text{ K}\right]$$

$$= 10\text{ bar}\left[\frac{0.287\,\dfrac{\text{kJ}}{\text{kg K}} \times 288\text{ K}}{1\text{ bar}}\right]\frac{14}{15}$$

or $$125.43 + 0.5829T_4 - 671.5 = 771.5 \qquad (T_4 \text{ in K})$$

or $$T_4 = \frac{1317.53}{0.5829} = 2260\text{ K} \quad (1987\,^\circ\text{C}).$$

$$\text{work ratio } r_{\text{w}} = \frac{w_{\text{net}}}{w_{+\text{ve}}}$$

and $$w_{\text{net}} = \text{m.e.p.}\left(\frac{RT_1}{p_1}\right)\left(1 - \frac{1}{15}\right) = \frac{1000\,\dfrac{\text{kN}}{\text{m}^2} \times 0.287\,\dfrac{\text{kJ}}{\text{kg K}} \times 288\text{ K} \times 14}{100\,\dfrac{\text{kN}}{\text{m}^2} \times 15}$$

$$= 771.5\,\frac{\text{kJ}}{\text{kg}}.$$

Now $$\left|w_{-\text{ve}}\right| = \frac{R(T_2 - T_1)}{\gamma - 1} = \frac{0.287\,\dfrac{\text{kJ}}{\text{kg K}}}{0.4}\,(850.8 - 288)\text{ K} = 403.8\,\frac{\text{kJ}}{\text{kg}}.$$

Thus $$w_{+\text{ve}} = w_{\text{net}} + \left|w_{-\text{ve}}\right| = 771.5 + 403.8 = 1175.3\,\frac{\text{kJ}}{\text{kg}}.$$

and $$r_{\text{w}} = \frac{771.5}{1175.3} = 0.656.$$

Example 11.7

An air standard cycle for an internal-combustion engine consists of the following processes:

(a) isentropic compression from 1 bar, 300 K, through a compression ratio of 6 : 1;
(b) heat addition at constant volume of 2000 kJ/kg;

(c) isentropic expansion to the initial volume;
(d) heat rejection at constant volume:

Sketch the cycle on a p–v diagram and calculate the ideal efficiency, mean effective pressure and peak pressure.

Answer

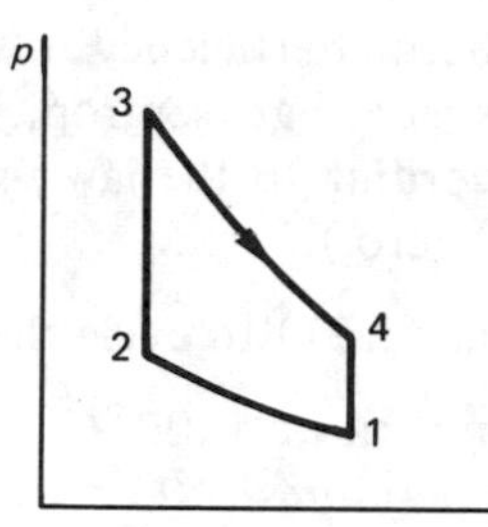

$$T_2 = T_1 \left(\frac{v_1}{v_2}\right)^{\gamma-1} = 300 \text{ K } (6)^{0.4} = 614.3 \text{ K.}$$

$$p_2 = p_1 \left(\frac{v_1}{v_2}\right)^{\gamma} = 1 \text{ bar } (6)^{1.4} = 12.29 \text{ bar.}$$

$$T_3 = T_2 + \frac{{}_2q_3}{c_v} = 614.3 \text{ K} + \frac{2000 \dfrac{\text{kJ}}{\text{kg}}}{0.718 \dfrac{\text{kJ}}{\text{kg K}}} = 3399.8 \text{ K.}$$

$$T_4 = T_3 \left(\frac{v_3}{v_4}\right)^{\gamma-1} = 3399.8 \text{ K } (\tfrac{1}{6})^{0.4} = 1660.3 \text{ K.}$$

$$w_{\text{net}} = {}_2q_3 + {}_4q_1 \qquad ({}_3q_4 = {}_1q_2 = 0)$$

$$= 2000 \frac{\text{kJ}}{\text{kg}} + (e_1 - e_4) \qquad ({}_4w_1 = 0)$$

$$= 2000 \frac{\text{kJ}}{\text{kg}} + c_v (T_1 - T_4);$$

$$\therefore w_{\text{net}} = 2000 \frac{\text{kJ}}{\text{kg}} + 0.718 \frac{\text{kJ}}{\text{kg K}} (300 - 1660.3) \text{ K} = 1023.3 \frac{\text{kJ}}{\text{kg}}.$$

$$\eta_{\text{th}} = \frac{w_{\text{net}}}{q_{+\text{ve}}} = \frac{1023.3}{2000} = 0.512.$$

$$v_1 - v_2 = v_1 \left(1 - \frac{v_2}{v_1}\right) = \frac{RT_1}{p_1} \left(1 - \frac{v_2}{v_1}\right) = \frac{0.287 \dfrac{\text{kJ}}{\text{kg K}} \times 300 \text{ K}}{100 \dfrac{\text{kN}}{\text{m}^2}} (1 - \tfrac{1}{6}) = 0.7175 \frac{\text{m}^3}{\text{kg}}.$$

$$\text{m.e.p.} = \frac{w_{\text{net}}}{v_1 - v_2} = \frac{1023.3 \dfrac{\text{kJ}}{\text{kg}}}{0.7175 \dfrac{\text{m}^3}{\text{kg}}} = 1426.2 \frac{\text{kN}}{\text{m}^2} = 14.26 \text{ bar.}$$

$$p_3 = p_2 \left(\frac{T_3}{T_2}\right) = 12.29 \text{ bar} \left(\frac{3399.8}{614.3}\right) = 68.02 \text{ bar.}$$

Exercises

1 A cycle consists of isochoric compression from 1 bar, 288 K, to 4 bar followed by isentropic expansion to 1 bar followed by isobaric cooling to the initial state.
Calculate the thermal efficiency, m.e.p. and work ratio. Air is the working fluid.
(0.211, 0.937 bar, 0.483)

2 Find the thermal efficiency, m.e.p. and work ratio for a cycle similar to that in example 2 except that isentropic expansion is replaced by reversible polytropic expansion according to the law $pv^{1.25}$ = constant. (Hint: the heat transfer in expansion is not zero.) (0.51, 12.38 bar, 0.77)

3 In an air standard Otto cycle the following data are given:

Maximum temperature 1400 °C
Minimum temperature 15 °C
Energy supplied 800 kJ/kg

Calculate the compression ratio, thermal efficiency, ratio of maximum to minimum pressures and work ratio. (5.24, 0.484, 30.43, 0.666.)

12 Heat Transfer

12.1 Introduction

Heat transfer is not thermodynamics. It may, however, be regarded as a satellite subject as the diagram tries to show.

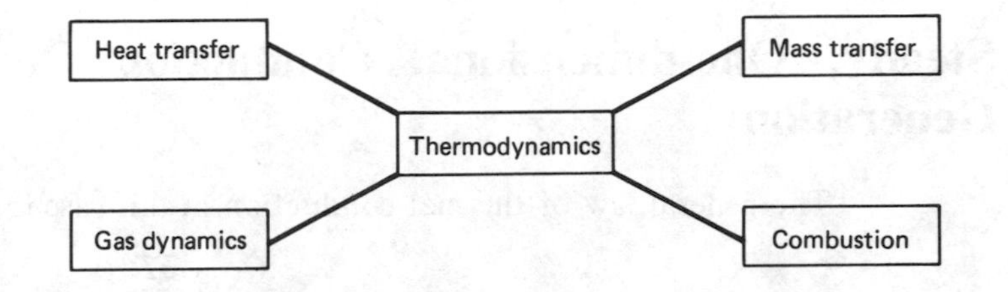

Thermodynamics as a subject depends for its development on the first and second laws (or more strictly axioms) of thermodynamics.

Heat transfer has its own quite distinctive laws of conduction, convection and radiation.

The link between the subject of thermodynamics and that of heat transfer arises because heat transfer appears quite logically when dealing with systems in equilibrium.

Heat transfer can take place only when there is a temperature gradient between the system and its surroundings to cause it. This gradient of temperature must be clearly distinguished from any temperature gradients which may exist within the thermodynamic system itself.

The science of heat transfer not merely seeks to explain how heat is transferred but also shows how a prediction can be made of *the rate at which* heat will transfer under certain specified conditions. That is, it introduces the quantity $\dot{Q}$ $(\mathrm{d}Q/\mathrm{d}t)$ from the point of view of experimentally determined evidence.

There are three modes of heat transfer which may exist separately or collectively in a given case.

(a) Thermal Conduction

This is by means of elastic impacts between molecules without appreciable displacement.

Energy transfer occurs from atoms of high vibrational energy to those of lower vibrational energy.

(b) Thermal Convection

Two kinds need identification.

(a) *Natural convection:* temperature gradients lead to density gradients which in turn lead to fluid buoyancy forces which cause bodily movement of the fluid and thus bodily energy transfer.

(b) *Forced convection:* bulk or mass movement of fluid and its energy is caused by an external agency such as a pump or fan with a temperature gradient between a fluid and a solid surface.

(c) Thermal Radiation

This is energy transfer by temperature-excited electromagnetic waves in the range 0.8 micron $< \lambda <$ 400 micron.

We consider one or two examples of conduction in this chapter together with some simple examples of convection but meaningful examples on radiation are considered to be beyond the scope of a first-year tutorial book like this.

12.2 Steady, One-dimensional Conduction (No Internal Energy Generation)

The general law of thermal conduction in this case is given by Fourier as follows:

$$\frac{\dot{Q}}{A} \propto \frac{\partial T}{\partial x}$$

where $\dot{Q}$ is the rate of heat transfer (kW),
∂T is the temperature gradient over length ∂x of the solid and
A is the area perpendicular to the direction of heat flow.

Rewriting this we can say

$$\dot{Q}'' = \frac{\partial T}{\partial x}$$

where the double dash sign refers to per unit area (single dash for unit length and triple dash for unit volume).

Inserting the constant of proportionality we get:

$$\dot{Q}'' = -k \frac{\partial T}{\partial x}.$$

Notes

(a) k is the thermal conductivity (which has dimensions given $\dot{Q}/[A(\partial T/\partial x)]$ W/(m K)).

(b) The negative sign in the equation arises because the heat transfers in the sense of decreasing temperature (in accordance with the second law of thermodynamics).

(c) Partial differential coefficients are quoted above to cover the more general case of unsteady conduction. However, since we are concerned here only with steady conduction we can rewrite the Fourier law as follows:

$$\dot{Q}'' = -k \frac{dT}{dx}.$$

(Note that k can be a variable in practice but in the examples that follow it is always taken as constant.)

The application of one-dimensional Fourier to certain simple cases of conduction heat transfer will now be considered.

Wherever possible, reference will be made for values of solid properties in this chapter from the table given on page 150.

12.3 Worked Example

Example 12.1

A cold room has one wall, 4.7 m by 2.3 m, made of brick 115 mm thick on the inside, a layer of cork 75 mm thick and wood 25 mm thick on the outside.

Estimate the heat transfer through the wall in 24 hours if the internal and external surface temperatures are −2 °C and 65 °C respectively. Also, calculate the temperatures of the interfaces between the brick and cork and cork and wood.

Answer

Conductivities (W/m K): brick, 0.9; cork, 0.04; wood, 0.17.

The figure gives a diagrammatic representation of the temperature distribution through the composite wall.

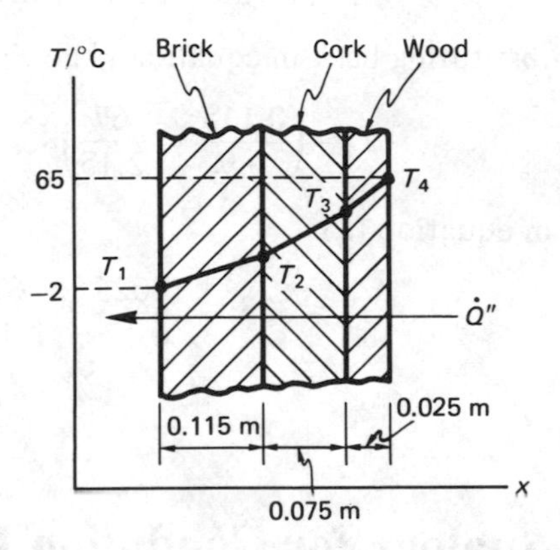

The first thing to realise is that one-dimensional Fourier can be applied to each of the three elements of the composite wall in turn:

Brick: $$\dot{Q}'' = \frac{k_B (T_2 - 271 \text{ K})}{x_B}.$$

(T_1 = 2 °C = 271 K and the sign of ΔT has now been changed to give ($T_2 - T_1$) inside bracket and a + sign outside!)

Cork: $$\dot{Q}'' = \frac{k_C (T_3 - T_2)}{x_C}.$$

Wood: $$\dot{Q}'' = \frac{k_W (338 \text{ K} - T_2)}{x_W}.$$

Thus the same $\dot{Q}''$ passes through each of the three elements of the wall in steady flow in turn – a very important feature of this type of problem which leads to a useful electrical analogy as shown at the end of this solution.

$$\text{Thus } \dot{Q}'' = \frac{0.9}{0.115} \frac{\text{W}}{\text{m}^2 \text{ K}} (T_1 - 271)\text{K} = \frac{0.04}{0.075} \frac{\text{W}}{\text{m}^2 \text{ K}} (T_2 - T_1)\text{K}$$

$$= \frac{0.17}{0.025} \frac{\text{W}}{\text{m}^2 \text{ K}} (338 - T_2)\text{K}$$

or $$(T_1 - 271)\text{K} = \frac{0.115}{0.9} \dot{Q}'' \frac{\text{m}^2 \text{ K}}{\text{W}} \quad \ldots (1)$$

and $$(T_2 - T_1)\text{K} = \frac{0.075}{0.04} \dot{Q}'' \frac{\text{m}^2 \text{ K}}{\text{W}} \quad \ldots (2)$$

and $$(338 - T_2)\,\mathrm{K} = \frac{0.025}{0.17}\,\dot{Q}\,\frac{\mathrm{m^2\ K}}{\mathrm{W}} \qquad \ldots (3)$$

Adding equations (1), (2) and (3) gives

$$(338 - 271)\,\mathrm{K} = \left(\frac{0.115}{0.9} + \frac{0.075}{0.04} + \frac{0.025}{0.17}\right)\dot{Q}''\,\frac{\mathrm{m^2\ K}}{\mathrm{W}}.$$

Thus $$\dot{Q}'' = \frac{67}{0.128 + 1.875 + 0.147}\,\frac{\mathrm{W}}{\mathrm{m^2}} = \frac{67}{2.15}\,\frac{\mathrm{W}}{\mathrm{m^2}}$$

and $$\dot{Q} = \dot{Q}'' \times A = \frac{67}{2.15}\,\frac{\mathrm{W}}{\mathrm{m^2}} \times 4.7\ \mathrm{m} \times 2.3\ \mathrm{m}$$

and $$Q = \dot{Q} \times t = \frac{67 \times 4.7 \times 2.3}{2.15}\,\mathrm{W} \times 24\ \mathrm{h}\left[\frac{\mathrm{kJ}}{10^3\ \mathrm{W\ s}}\right]\left[\frac{3600\ \mathrm{s}}{\mathrm{h}}\right]$$

$$Q = 291\,06\ \mathrm{kJ} = 29.11\ \mathrm{MJ}.$$

Substituting back in equation (1) we get

$$T_1 = \left(\frac{0.115}{0.9} \times \frac{67}{2.15}\right)\mathrm{K} + 271\ \mathrm{K} = 275\ \mathrm{K} \quad (= 2\,°\mathrm{C}),$$

and in equation (3)

$$T_2 = 338 - \frac{0.025}{0.17} \times \frac{67}{2.15} = 333.4\ \mathrm{K} \quad (60.4\,°\mathrm{C}).$$

12.4 Electrical Analogy for Condution Heat Transfer

As will be clear from the foregoing, the driving force for heat transfer is a temperature gradient. This may be considered to be analogous to potential difference in an electrical circuit which causes current flow against an electrical resistance. The equivalent quantity to current flow is of course the heat transfer and the equivalent quantity to electrical resistance will be the bracketed quantity ($\Delta x/k$).

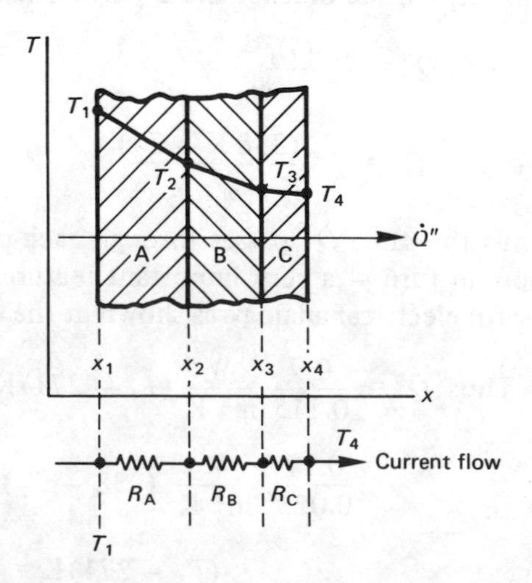

Thus for a composite plane wall as in the above example (see diagram) we can write in general terms:

$$\dot{Q}'' = \frac{-k_A (T_2 - T_1)}{(x_2 - x_1)} = \frac{-k_B (T_3 - T_2)}{x_3 - x_2} = \frac{-k_C (T_4 - T_2)}{(x_4 - x_3)}$$

$$= \frac{(T_1 - T_4)}{\frac{(x_2 - x_1)}{k_A} + \frac{(x_3 - x_2)}{k_B} + \frac{(x_4 - x_3)}{k_C}}$$

$$= \frac{\Delta T}{\Sigma R_{\text{thermal}}} \quad \text{where } R_{\text{thermal}} = \frac{\Delta x}{k}.$$

The electrical analogy may be extended to more complex configurations as below.

12.5 Series and Parallel Flow in Heat Conduction

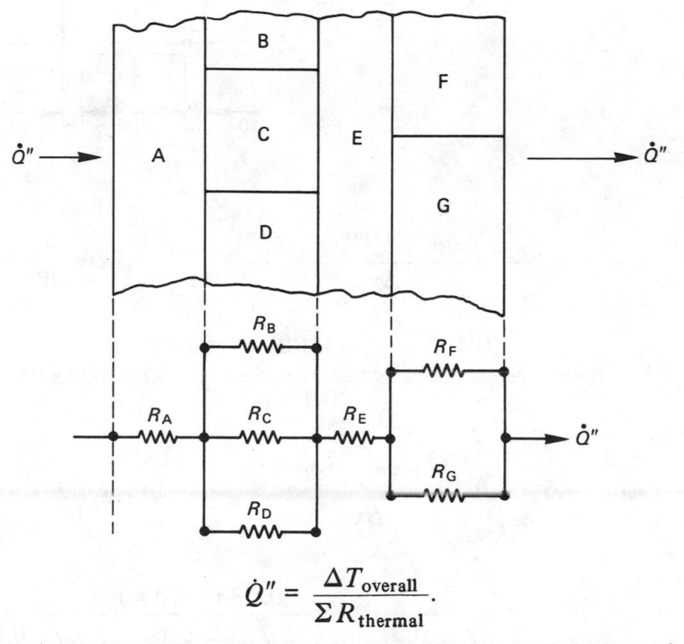

$$\dot{Q}'' = \frac{\Delta T_{\text{overall}}}{\Sigma R_{\text{thermal}}}.$$

This topic is best dealt with by means of the following worked example.

12.6 Worked Example

Example 12.2

A special insulating board is made up as shown in the diagram (all dimensions in mm; not to scale). The outer panels and dividing partitions have a thermal con-

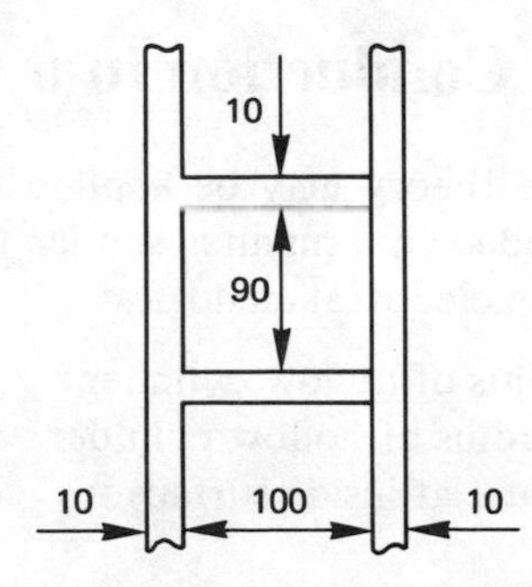

ductivity of 0.105 W/(m K) while the horizontal compartments are filled with granular material having a conductivity of 0.032 W/(m K). If the mean outside temperatures are 70 °C and 20 °C, determine the heat transfer rate per unit area of the board. Assume one-dimensional heat flow normal to the plane of the board.

Answer

Consider the unit width of the board into the page (all dimensions in mm; not to scale):

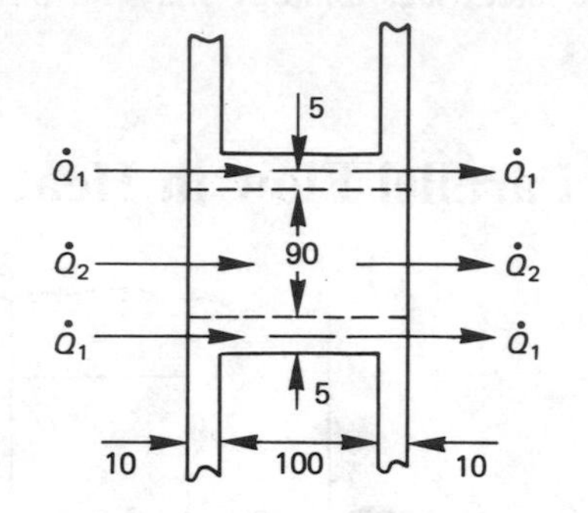

$$\dot{Q}_1' = \frac{k_1 \dfrac{A_1}{W}(T_{\text{upper}} - T_{\text{lower}})}{\Delta x} \qquad \text{where } \frac{A_1}{W} = \text{area per unit width for narrow passage}$$

$$= \frac{0.105 \dfrac{\text{W}}{\text{m K}} \times 0.005 \text{ m } (50 \text{ K})}{0.12 \text{ m}} = 0.218\,75 \frac{\text{W}}{\text{m}}.$$

$$\dot{Q}_2' = \frac{\dfrac{A_2}{W}(T_{\text{upper}} - T_{\text{lower}})}{\Sigma \dfrac{\Delta x}{k}} \qquad \text{where } \frac{A_2}{W} = \text{area per unit width for wide passage}$$

$$= \frac{0.09 \text{ m } (50 \text{ K})}{\left(\dfrac{0.01 \text{ m}}{0.105 \dfrac{\text{W}}{\text{m K}}}\right) + \left(\dfrac{0.1 \text{ m}}{0.032 \dfrac{\text{W}}{\text{m K}}}\right) + \left(\dfrac{0.01 \text{ m}}{0.105 \dfrac{\text{W}}{\text{m K}}}\right)} = 1.3573 \frac{\text{W}}{\text{m}}.$$

Total $\qquad \dot{Q}' = 2\dot{Q}_1 + \dot{Q}_2 = 2\,(0.21875) + 1.3573 = 1.795 \dfrac{\text{W}}{\text{m}}$

and $\qquad \dot{Q}'' = \dfrac{\dot{Q}'}{W} = \dfrac{1.795 \dfrac{\text{W}}{\text{m}}}{1 \text{ m}} = 1.795 \dfrac{\text{W}}{\text{m}^2}.$

12.7 Steady Radial Conduction in a Cylinder

The above theory may be applied to radial steady conduction in a single cylinder and extended in a manner similar to the above to a composite cylinder with corresponding electrical analogies:

Inside radius of hollow cylinder = r_1.
Outside radius of hollow cylinder = r_2.
Temperature at inside surface = T_1 ($> T_2$).

Temperature at outside surface = T_2.
Temperature at radius $r = T$.
Temperature at radius $r + \mathrm{d}r = T + \mathrm{d}T$.

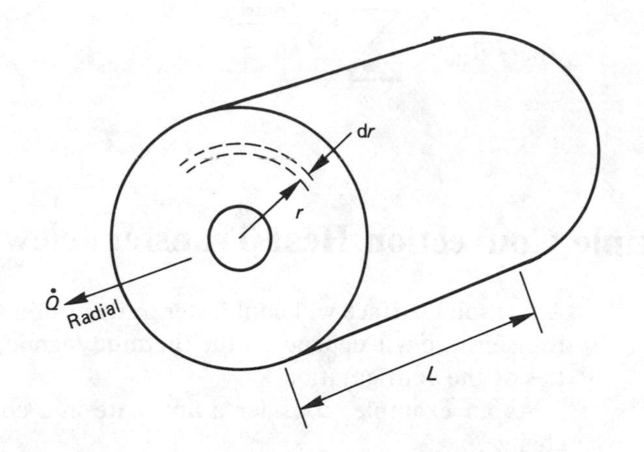

$$\dot{Q} = -kA \frac{\mathrm{d}T}{\mathrm{d}r} \quad \text{(Fourier)}$$

or $$\dot{Q} = -2\pi krL \frac{\mathrm{d}T}{\mathrm{d}r}. \qquad (A \text{ is } A_{\text{radial}} \text{ not } A_{\text{x-section}}\,!)$$

Integrating between $r = r_1;\ T = T_1$

and $r = r_2;\ T = T_2$,

$$\dot{Q} = \frac{2\pi kL\,(T_1 - T_2)}{\ln \frac{r_2}{r_1}} = \frac{\Delta T}{R_{\text{thermal}}} \qquad \text{where } R_{\text{thermal}} = \frac{\ln \frac{r_2}{r_1}}{2\pi kL}$$

i.e. the electrical analogy for a single cylinder is

T_1 —R_{th}— T_2 → $\dot{Q}$ $$R_{th} = \frac{\ln \frac{r_2}{r_1}}{2\pi kL}.$$

Extension of the above to a multiple cylinder:

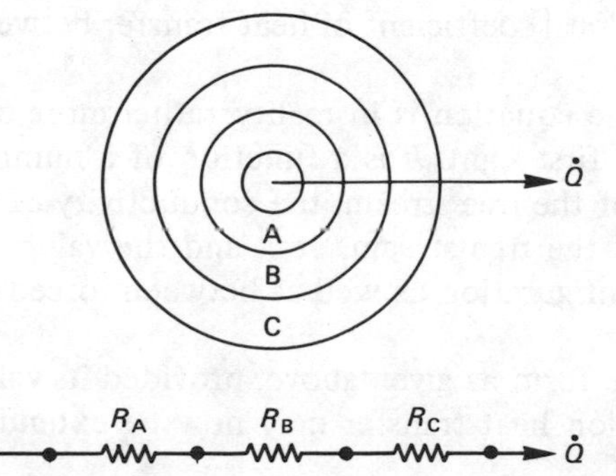

Length L into the page.
Radius at inside of A is r_1 (temp. T_1).
Radius at inside of B is r_2.
Radius at inside of C is r_3.
Radius at outside of C is r_4 (temp. T_4).

R_A — R_B — R_C → $\dot{Q}$ Electrical equivalent:

$$\dot{Q} = \frac{2\pi k L (T_1 - T_4)}{\dfrac{\ln \frac{r_2}{r_1}}{k_A} + \dfrac{\ln \frac{r_3}{r_2}}{k_B} + \dfrac{\ln \frac{r_4}{r_3}}{k_C}} = \frac{\Delta T}{R_{th}}$$

$$\text{where } R_{th} = \sum \frac{\ln \frac{r_{outer}}{r_{inner}}}{k}.$$

12.8 Simple Convection Heat Transfer (Newton's Law)

A hot solid surface will cool faster in front of a fan than when in still air. The heat transfer rate will depend on the thermodynamic, transport and mechanical properties of the configuration.

As an example, consider a hot plate in a colder atmosphere showing velocity change.

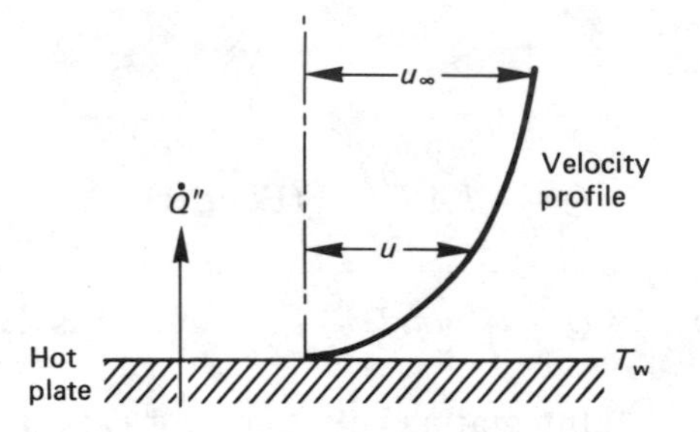

At the wall surface $u = 0$, and heat transfer is effected by thermal conduction from the wall through a stagnant boundary layer and thence by convection to the free stream outside.

12.9 Newton's Law of Cooling

$$\dot{Q}'' = h(T_w - T_\infty)$$

where T_∞ is the free-stream temperature.
h is called the surface or local coefficient of heat transfer between the solid surface and the free stream.

This deceptively simple equation is in reality rather more complex than would appear to be the case at first sight. h is a function of a number of different variables, e.g. the velocity of the free stream, the conductivity of the free stream, the specific heat capacity of the free stream, etc., and the value of h differs for each different geometrical configuration as well as between forced and natural convection.

However, in its simple form as given above, provided its value can be specified, the analysis of conduction heat transfer may now be extended as the following examples demonstrate.

12.10 Worked Examples

Example 12.3

Steam for space-heating purposes on a factory site is supplied from the boiler house to a remote building through a 90 m length of lagged pipe carried above ground level on pylon supports.

Steam enters the pipe at 3.5 bar, 0.98 dry. The pipe has an inside diameter of 0.15 m, a wall thickness of 10 mm and a conductivity of 48 W/(m K). It is lagged with a 50 mm thickness of insulating material having a conductivity of 0.87 W/(m K).

Heat transfer coefficients between the steam and pipe and from the lagging to the atmosphere are 2.84 and 34.1 kW/(m^2 K) respectively.

If the steam mass flow rate is 2.5 kg/s and the atmospheric temperature is 0 °C, calculate the steam quality at exit ignoring pipe friction.

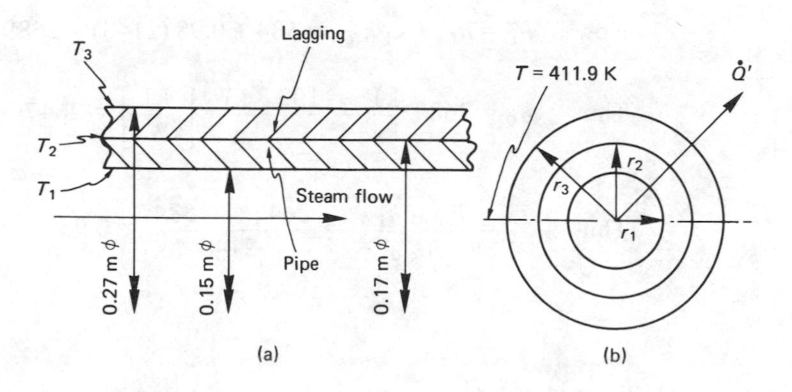

Answer

$\Delta p = 0$ (frictionless).

For wet steam at exit, T_{steam} at 3.5 bar = 138.9 °C
= 411.9 K (constant). (p. 4 of tables)

For pipe, $$\dot{Q}' = \frac{2\pi k_p (T_1 - T_2)}{\ln \frac{r_2}{r_1}} = \frac{2\pi \times 48 \ \frac{\text{W}}{\text{m K}} (T_1 - T_2)\ \text{K}}{\ln \frac{0.17}{0.15}}.$$

For lagging, $$\dot{Q}' = \frac{2\pi k_L (T_2 - T_3)}{\ln \frac{r_3}{r_2}} = \frac{2\pi \times 0.87 \ \frac{\text{W}}{\text{m K}} (T_2 - T_3)\ \text{K}}{\ln \frac{0.27}{0.17}}.$$

At inner surface, $$\dot{Q}' = \pi D_1 m \times h_{\text{inner}} \frac{\text{kW}}{\text{m}^2\ \text{K}} (411.9 - T_1)\text{K}.$$

At outer surface, $$\dot{Q}' = \pi D_3 m \times h_{\text{outer}} \frac{\text{kW}}{\text{m}^2\ \text{K}} (T_3 - 273)\ \text{K}.$$

Rearranging the above,

$$411.9 - T_1 = \frac{\dot{Q}'}{\pi D_1 h_{\text{inner}}}; \qquad T_2 - T_3 = \frac{Q' \ln \frac{r_3}{r_2}}{2\pi k_1};$$

$$T_1 - T_2 = \frac{\dot{Q}' \ln \frac{r_2}{r_1}}{2\pi k_p}; \qquad T_3 - 273 = \frac{\dot{Q}'}{\pi D_3 h_{\text{outer}}}.$$

Adding above, we get

$$(411.9 - 273)\ \text{K} = \dot{Q}' \left[\frac{1}{0.15\pi \times 2840} + \frac{\ln \dfrac{0.17}{0.15}}{2\pi \times 48} + \frac{\ln \dfrac{0.27}{0.17}}{2\pi \times 0.87} + \frac{1}{0.27\pi \times 34.1} \right] \frac{\text{m K}}{\text{W}}$$

$$138.9 = \dot{Q}' (0.000\,75 + 0.000\,41 + 0.08463 + 0.0346)\ \frac{\text{m}}{\text{W}} = \dot{Q}' \times 0.120\,36\ \frac{\text{m}}{\text{W}}.$$

$$\dot{Q}' = \frac{138.9}{0.120\,36}\ \frac{\text{W}}{\text{m}} \quad \text{and} \quad \dot{Q} = \frac{138.9}{0.120\,36}\ \frac{\text{W}}{\text{m}} \times 90\ \text{m} = 103.8\ \text{kW}.$$

For incoming and outgoing specific enthalpies of steam of h_i and h_o, $\dot{Q} = \dot{m}(h_o - h_i)$ from s.f.e.e. and $\dot{Q}$ is rejected by steam.

Thus $\quad -103.8\ \text{kW} = 2.5\ \dfrac{\text{kg}}{\text{s}}\ (h_o - h_i).$

Now $\quad h_i = h_{f,i} + x_i h_{fg,i} = 584 + 0.98\,(2148) = 2689\ \dfrac{\text{kJ}}{\text{kg}}.$ (p. 4)

Thus $\quad h_o = 2689\ \dfrac{\text{kJ}}{\text{kg}} - \dfrac{103.8\ \text{kW}}{2.5\ \dfrac{\text{kg}}{\text{s}}} \left[\dfrac{\text{kJ}}{\text{kW s}}\right] = 2647.5\ \dfrac{\text{kJ}}{\text{kg}}$

Thus $\quad x_o = \dfrac{h_o - h_{f,o}}{h_{fg,o}} = \dfrac{2647.5 - 584}{2148} = 0.96.$

Example 12.4

A cylindrical conductor of radius r, which is at a uniform temperature, is covered with electrical insulation of thermal conductivity k. The heat transfer coefficient between the surface of the insulation and the surrounding air is h. What is the outside radius of the insulation for the heat transfer from the conductor to be a maximum for the given conductor and air temperatures?

Show that the insulation will act as thermal insulation only if its outside radius is greater than the value of r_2 given by

$$r_1/r_2 + (hr_1/k)\ln(r_2/r_1) = 1.$$

If r_1 is 3 mm, h is 10 W/(m^2 K) and k is 0.1 W/(m K), determine the value of r_2.

This is an important problem because the distinction is now drawn between the use of lagging material up to a certain radius to promote heat transfer from the inner conductor (as in an electrical cable to avoid burn-out) and extra lagging to reduce heat transfer where insulation against excessive heat loss is required.

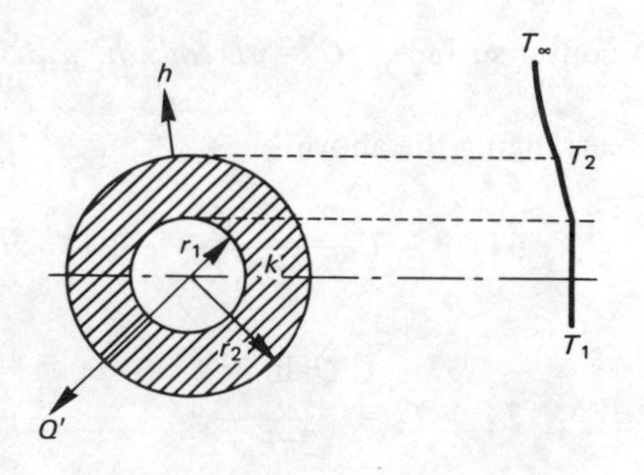

Answer

$$\dot{Q}' = -kA\,\frac{\mathrm{d}T}{\mathrm{d}r} = hA\,\Delta T \quad \text{from surface.}$$

$$\dot{Q}' = -k2\pi r \cdot \frac{\mathrm{d}T}{\mathrm{d}r}$$

$$\int_{r_1}^{r_2} \dot{Q}'\,\frac{\mathrm{d}r}{r} = -2\pi k \int_{T_1}^{T_2} \mathrm{d}T.$$

$$\dot{Q}' \ln\frac{r_2}{r_1} = -k2\pi(T_2 - T_1) = k2\pi(T_1 - T_2).$$

Thus $$\dot{Q}' = \frac{2\pi k(T_1 - T_2)}{\ln\frac{r_2}{r_1}} = hA_2(T_2 - T_\infty)$$

Now $$T_1 - T_2 = \frac{\dot{Q}' \ln\frac{r_2}{r_1}}{2\pi k}$$

and $$T_2 - T_\infty = \frac{\dot{Q}'}{hA_2} = \frac{\dot{Q}'}{2h\pi r_2}.$$

Adding $$T_1 - T_\infty = \dot{Q}'\left[\frac{\ln\frac{r_2}{r_1}}{2\pi k} + \frac{1}{2\pi h r_2}\right] \quad \text{(a constant } \Delta T\text{)}.$$

Thus $$\dot{Q}' = \frac{(T_1 - T_\infty)}{\frac{\ln\frac{r_2}{r_1}}{2\pi k} + \frac{1}{2\pi h r_2}}$$

or $$\dot{Q}' = (T_1 - T_\infty)\left[\frac{\ln\frac{r_2}{r_1}}{2\pi k} + \frac{1}{2\pi h r_2}\right]^{-1}.$$

Thus $$\frac{\mathrm{d}\dot{Q}'}{\mathrm{d}r_2} = (T_1 - T_\infty)(-1)\left[\frac{\ln\frac{r_2}{r_1}}{2\pi k} + \frac{1}{2\pi h r_2}\right]^{-2}\left[\frac{r_1\left(\frac{1}{r_1}\right)}{r_2\,2\pi k} + \frac{1}{2\pi h}\left(-\frac{1}{r_2^2}\right)\right]$$

$$= 0 \text{ for stationary value.}$$

Thus $$\frac{1}{2\pi r_2 k} = \frac{1}{2\pi h r_2^2}$$

or $$r_2 = \frac{k}{h}.$$

It is left to you to show that $\dfrac{\mathrm{d}^2\dot{Q}}{\mathrm{d}r_2^2}$ is negative with the substitution of this value.

Now $$\dot{Q}' = \frac{(T_1 - T_\infty)}{\frac{\ln\frac{r_2}{r_1}}{2\pi k} + \frac{1}{2\pi h r_2}}$$

and, since $2\pi(T_1 - T_\infty)$ is fixed, then, for $\dot{Q}'$ to decrease (i.e. thermal insulation),

$$\frac{\ln \frac{r_2}{r_1}}{k} + \frac{1}{hr_2}$$ must increase or the limit is given by

$$\frac{\ln \frac{r_2}{r_1}}{k} + \frac{1}{hr_2} = 1 \quad \text{or} \quad \frac{hr_1}{k} \ln \frac{r_2}{r_1} + \frac{hr_1}{hr_2} = 1 \quad \text{or} \quad \frac{hr_1}{k} \ln \frac{r_2}{r_2} + \frac{r_1}{r_2} = 1 \quad \text{in limit.}$$

For $r_1 = 3$ mm, $h = 10 \dfrac{\text{W}}{\text{m}^2\ \text{K}}$, $k = 0.1 \dfrac{\text{W}}{\text{m K}}$:

$$\frac{10 \dfrac{\text{W}}{\text{m}^2\ \text{K}} (0.003\ \text{m})}{0.1 \dfrac{\text{W}}{\text{m K}}} \ln \frac{r_2}{0.003} + \frac{0.003}{r_2} = 1.$$

Thus
$$100\,(0.003) \ln \frac{r_2}{0.003} + \frac{0.003}{r_2} = 1$$

for which the solution is $r_2 = 0.0734$ m = 73.4 mm.

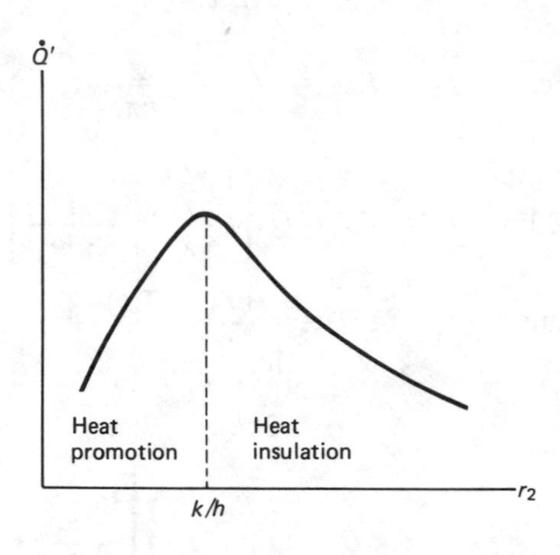

12.11 Unsteady-state Heat Conduction—No Internal Energy Generation

This class of problem deals with the starting up and running down periods involved with the switching on and switching off of equipment that ultimately runs in a steady-state condition such as that treated previously in this chapter.

Many heat transfer problems in unsteady flow can readily be solved by assuming that the internal resistance to heat flow is negligible compared with the external resistance between the solid surface and the surrounding medium so that the latter effectively controls the heat transfer process.

That is, the temperature throughout the system at any one time is uniform.

One application of this is on the heating and cooling of small bodies, e.g. the heating of a billet in a furnace.

The assumptions are:

(a) the billet has a high value of k (i.e. low R_{th}),
(b) the introduction to the furnace gives an instantaneous step change in temperature from environment T_0 to the furance T_f.

Nomenclature: Temperature on entry to furnace T_0
Furnace or bulk fluid temperature T_f
Billet temperature after time t in furnace T
Surface heat transfer coefficient h
Billet specific-heat capacity c
Billet volume V
Billet density ρ
Billet surface area A_S

Assuming that the temperature change of the billet in the furnace after a time dt is dT, then for time increasing,

$$\dot{Q}\,dt = \rho Vc\,dT = mc\,dT,$$

i.e.

$$\dot{Q} = mc\frac{dT}{dt}. \qquad \ldots (1)$$

But

$$\dot{Q} = hA_S(T_f - T) \qquad \text{(at the surface of billet).}$$

Thus $$\frac{dT}{T_f - T} = \frac{hA_S}{mc}\,dt,$$

i.e. $$\frac{d(T_f - T)}{T_f - T} = -\frac{hA_S}{mc}\,dt. \qquad \ldots (3)$$

(*Note:* T_f is a constant and change sign on both sides.)

Limits: $\left.\begin{matrix} t = 0 \\ T = T_0 \end{matrix}\right\}$ $\left.\begin{matrix} t = t \\ T = T \end{matrix}\right\}$.

Integrating, $$\frac{T_f - T}{T_f - T_0} = \exp -\left(\frac{hA_S}{mc}t\right) = \exp -\frac{t}{t_0} \qquad \left(t_0 = \frac{mc}{hA_S}\right).$$

Note that t_0 is defined as the *time constant* for the circuit.

Electrical Analogy

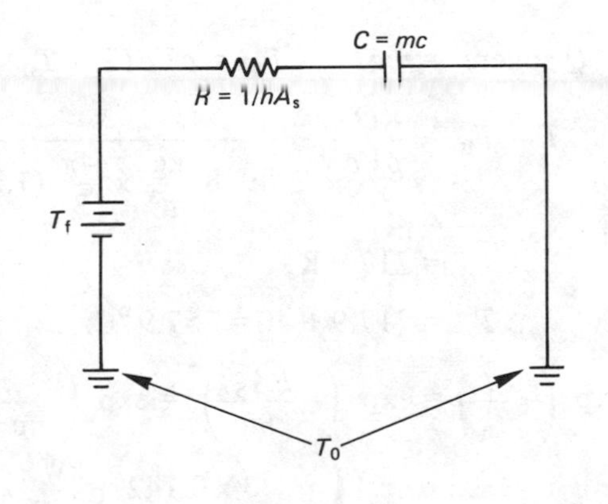

12.12 Worked Examples

Example 12.5

Calculate the time need for a spherical casting made of aluminium to be heated in a furnace to 510 °C by gases at 1200 °C if the casting is put into the oven at 16 °C. The casting diameter is 0.5 m and its average h (surface) is 0.085 kW/(m² K).

Answer

Time constant

$$t_0 = \frac{mc}{hA_S} = \frac{\rho Vc}{hA_S} = \frac{\rho \frac{4}{3}\pi r^3 c}{h4\pi r^2} = \frac{\rho rc}{3h}$$

$$= \frac{2707 \frac{\text{kg}}{\text{m}^3} \times 0.25 \text{ m} \times 0.871 \frac{\text{kJ}}{\text{kg K}}}{3 \times 0.085 \frac{\text{kW}}{\text{m}^2 \text{ K}}} \left[\frac{\text{kW s}}{\text{kJ}}\right] = 2312 \text{ s}.$$

$$\frac{T - T_f}{T_0 - T_f} = \frac{510 - 1200}{16 - 1200} = 0.583.$$

$$\frac{t}{t_0} = -\ln\left(\frac{T - T_f}{T_0 - T_f}\right) = -\ln 0.583 = 0.54.$$

$$t = 0.54 \times 2312 \text{ s} \left[\frac{\text{min}}{60 \text{ s}}\right] = 20.8 \text{ min}.$$

Example 12.6

A gas-fired preheater for a wind tunnel facility is to absorb 200 000 kJ during a 3 minute heating period.

The preheater contains 10 000 solid copper spheres 2.5 cm in diameter at an initial uniform temperature of 40 °C.

Neglecting the internal thermal resistance of the spheres, and assuming a uniform h of 0.142 kW/(m² K), show that the gas temperature must be at least 900 °C.

Answer

$$Q \text{ sphere} = mc(T - T_0) = \rho Vc(T - T_0);$$

$$T - T_0 = \frac{Q}{\rho Vc} = \frac{20 \text{ kJ}}{8938 \frac{\text{kg}}{\text{m}^3} \times \frac{4\pi}{3}(1.25^3 \text{ cm}^3)\left[\frac{\text{m}^3}{10^6 \text{ cm}^3}\right] \times 0.381 \frac{\text{kJ}}{\text{kg K}}}$$

$$= 717.9 \text{ K},$$

$$\therefore T = 717.9 + 40 = 757.9 \text{ °C}.$$

$$\exp\left(-\frac{t}{t_0}\right) = \exp\left(-\frac{hA_S t}{mc}\right) = \exp\left(-\frac{h4\pi r^2 t}{\rho \frac{4}{3}\pi r^3 c}\right) = \exp\left(-\frac{3ht}{r\rho c}\right)$$

$$\exp\left(-\frac{t}{t_0}\right) = \exp\left\{-\frac{3 \times 0.142 \frac{\text{kW}}{\text{m}^2 \text{ K}} \times 180 \text{ s}\left[\frac{\text{kJ}}{\text{kW s}}\right]}{0.0125 \text{ m} \times 8938 \frac{\text{kg}}{\text{m}^3} \times 0.381 \frac{\text{kJ}}{\text{kg K}}}\right\} = \exp(-1.8) = 0.165.$$

$$\frac{T - T_f}{T_0 - T_f} = 0.165 \quad \text{or} \quad 757.9 - T_f = 0.165(40 - T_f),$$

giving $T_f = \frac{751.3}{0.835} = 899.7 \text{ °C} \quad (\simeq 900 \text{ °C}).$

12.13 Unsteady-state Heat Conduction With Internal Energy Generation

This is the second of two classes of unsteady-state heat conduction problems, this time considering the inclusion of a constant internal-energy generation as for example in electrically heated windscreens.

Nomenclature:

$\dot{Q}_g$	= internal energy generation rate	(constant).
T_f	= bulk fluid temperature around body	(constant).
T_0	= initial body temperature at $t = 0$	(constant).
T	= body temperature at time t	(variable).
h	= convective h.t. coefficient at surface	(constant).
A_S	= body surface area	(constant).
m	= body mass	(constant).
c	= body specific-heat capacity	(constant).
T_∞	= body temperature at $t = \infty$	(constant).

Energy balance:

Internal energy generation = Gain of internal energy in body
+ Convective heat transfer from surface

or
$$\dot{Q}_g = mc\,\frac{dT}{dt} + hA_S(T - T_f)$$

or
$$\dot{Q}_g - \dot{Q}_S = mc\,\frac{dT}{dt} \qquad \text{(where } \dot{Q}_S = \text{heat transfer from surface).}$$

But
$$\dot{Q}_S = hA_S(T - T_f) = \frac{(T - T_f)}{R_S} \qquad (R_S = \text{surface thermal resistance).}$$

Thus
$$\int_0^t dt = mcR_S \int_{T_f}^{T} \frac{dT}{\dot{Q}_g R_S - (T - T_f)}.$$

Now
$$\dot{Q}_g R_S = T_\infty - T_f$$

(where T_∞ is body temperature when *fully* heated and gain of internal energy in body is now zero).

Thus
$$\int_0^t dt = mcR_S \int_{T_f}^{T} \frac{d(T - T_f)}{(T_\infty - T_f) - (T - T_f)}.$$

$$t = mcR_S\left\{-\ln\Big[(T_\infty - T_f) - (T - T_f)\Big]_{T_f}^{T}\right\}$$

$$\frac{t}{mcR_S} = \ln\left(\frac{T_\infty - T_f}{T_\infty - T}\right)$$

or
$$\exp\left(+\frac{t}{mcR_S}\right) = \frac{T_\infty - T_f}{T_\infty - T}$$

or
$$1 - \exp\left(-\frac{t}{mcR_S}\right) = 1 - \left(\frac{T_\infty - T}{T_\infty - T_f}\right) = \left(\frac{T_\infty - T_f}{T_\infty - T_f}\right) - \left(\frac{T_\infty - T}{T_\infty - T_f}\right)$$

or
$$1\left(-\exp - \frac{t}{t_0}\right) = \frac{T - T_f}{T_\infty - T_f} \qquad \text{where } t_0 = mcR_S.$$

Electrical Analogy

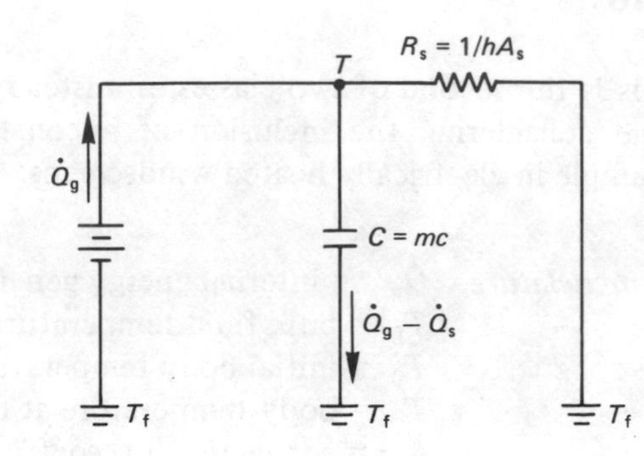

If $t = t_0$; $1 - \exp(-t/t_0) = 1 - 1/e = 0.6321$. The transient temperature change is depicted in the graph here.

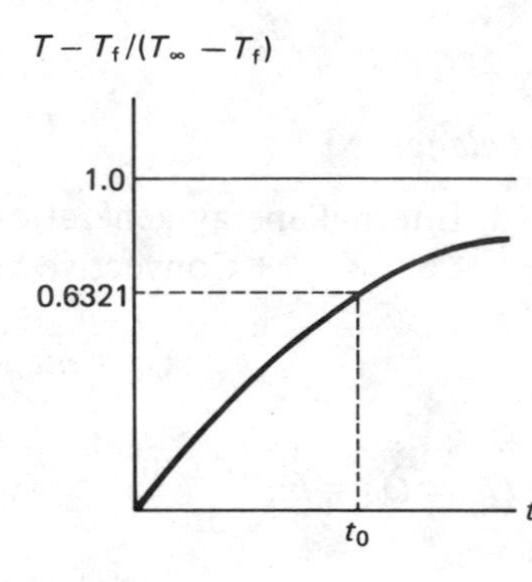

12.14 Worked Examples

Example 12.7

The power supply to an electrically heated windscreen panel is regulated to switch ON when the aircraft enters a region where ice is likely to form on the windscreen. The electrical input of 1500 W is obtained from a transformer of efficiency 0.97 which overheats in continuous use at sea level. The maximum permissible temperature of the windings is 176 °C. Estimate the maximum time for which de-icing may be used at sea level in the following circumstance:

Surroundings and intial windings temperature	−1.2 °C
Mass of core and windings	2.72 kg
Specific heat capacity of windings	0.419 kJ/(kg K)
Surface area of windings for heat transfer	0.0065 m²
Sea-level heat transfer coefficient	1.136 kW/(m² K)

Answer

$$T_0 - T_f = 0 \qquad (T_f = T_0 = 1.2\,°\text{C}).$$

$$t_0 = \frac{mc}{hA_s} = \frac{2.72\ \text{kg} \times 0.419\ \dfrac{\text{kJ}}{\text{kg K}}}{1.136 \times 10^{-2}\ \dfrac{\text{kW}}{\text{m K}} \times 0.0065\ \text{m}^2} = 15\,434\ \text{s} \quad (4.29\ \text{h}).$$

$$\frac{\dot{Q}_g}{hA_S} = \frac{1.5 \text{ kW}}{0.97} \times \frac{10^2}{1.136} \frac{\text{m}^2 \text{ K}}{\text{kW}} \times \frac{10^2}{0.65 \text{ m}^2} = 20\,942 \text{ K}.$$

Generally, $\dfrac{\dot{Q}_g}{hA_S}\left[1 - \exp\left(-\dfrac{t}{t_0}\right)\right] = T - T_f.$

Particularly, $\dfrac{\dot{Q}_g}{hA_S}\left[1 - \exp\left(-\dfrac{\hat{t}}{t_0}\right)\right] = \hat{T} - T$ $\qquad (\hat{T} = T_{maximum}).$

Thus $1 - \exp -\dfrac{\hat{t}}{t_0} = \dfrac{[176 - (-1.2)] \text{ K}}{20\,942 \text{ K}} = 0.008\,46.$

Thus $\exp \dfrac{\hat{t}}{t_0} = 1.0085$

or $\dfrac{\hat{t}}{t_0} = 0.0085$

and $\hat{t} = 0.0085 \times 15\,434 \text{ s}\left[\dfrac{\text{min}}{60 \text{ s}}\right] = 2.186 \text{ min}.$

Example 12.8

The starter of a d.c. motor is to be used for emergency braking, during which the power dissipated in the windings may be assumed to vary linearly from 100 kW to zero in 40 seconds. The mass of the windings is 22.7 kg, with surface area 0.743 m^2 and specific heat capacity of 0.377 kJ/(kg K). The surface heat transfer coefficient is $0.0454 \text{ kW/(m}^2 \text{ K)}$.

Show that the temperature of the windings is given by

$$T = a + b(t - t_0) + de^{-t/t_0}$$

where t is time, t_0 is the thermal time constant ($= mc/hA_S$) and a, b and d are constants.

Prove that the maximum temperature of the windings occurs approximately 40 seconds after the emergency when the surroundings are at 15 °C.

Answer

Slope of graph $= \dfrac{(0 - 100) \text{ kW}}{(40 - 0) \text{ s}} = -2.5 \dfrac{\text{kW}}{\text{s}}.$

$$\dot{Q}_g - \dot{Q}_{g,0} = -2.5 \frac{\text{kW}}{\text{s}} (t - 0)$$

$$\dot{Q}_g = \dot{Q}_{g,0} - 2.5 \frac{\text{kW}}{\text{s}} t.$$

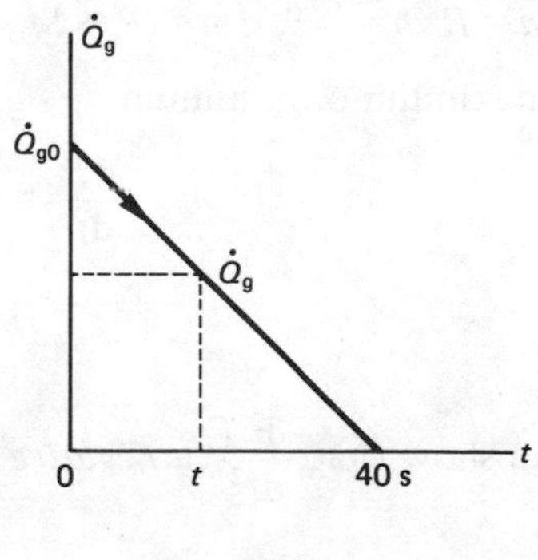

$\dot{Q}_g$

$\dot{Q}_{g0}$ $\qquad \dot{Q}_g$

0 $\qquad t \qquad$ 40 s $\qquad t$

Thus $\dot{Q}_{g,0} - 2.5 \dfrac{\text{kW}}{\text{s}} t = mc \dfrac{dT}{dt} + hA_S (T - T_f)$

i.e.

$$\frac{\dot{Q}_{g,0}}{hA_S} - \frac{2.5 \frac{\text{kW}}{\text{s}} t}{hA_S} = t_0 \frac{dT}{dt} + (T - T_f) \qquad \left(\text{where } t_0 = \frac{mc}{hA_S}\right)$$

or

$$R - St = t_0 \frac{dT}{dt} + T \qquad \left(\text{where } R = \frac{\dot{Q}_{g,0}}{hA_S}, \quad S = \frac{2.5 \frac{\text{kW}}{\text{s}}}{hA_S}\right)$$

This equation is of the form

$$\frac{dy}{dx} + Py = Q \qquad \text{(where } P \text{ and } Q \text{ are functions of } x \text{ or constants).}$$

Here $P = \text{constant} = \dfrac{1}{t_0}$ and the integrating factor is

$$e^{\int P\,dx} = e^{\int dt/t_0} = e^{t/t_0}$$

Now in the above, $\dfrac{dT}{dt} + \dfrac{T}{t_0} = \dfrac{R - St}{t_0}$,

and, multiplying through by $t_0 e^{t/t_0}$,

$$\underbrace{t_0 \frac{dT}{dt} e^{t/t_0} + Te^{t/t_0}}_{t_0 \frac{d}{dt}(Te^{t/t_0}),} = Re^{t/t_0} - Ste^{t/t_0}$$

i.e.

$$Te^{t/t_0} = \frac{1}{t_0} \int_0^t (Re^{t/t_0} - Ste^{t/t_0})\,dt$$

$$Te^{t/t_0} = \frac{1}{t_0} \left[Rt_0 e^{t/t_0}\right]_0^t - \frac{S}{t_0} \left[e^{t/t_0}(tt_0 - t_0^2)\right]_0^t$$

since $xe^{ax}\,dx = e^{ax}\left(\dfrac{x}{a} - \dfrac{1}{a^2}\right)$ and $a = \dfrac{1}{t_0}$.

Thus $\quad Te^{t/t_0} = R(e^{t/t_0} - 1) - S[e^{t/t_0}(t - t_0) - 1(0 - t_0)]$

or $\quad T = R - Re^{-t/t_0} - S(t - t_0) - St_0 e^{-t/t_0}$

or $\quad T = R - S(t - t_0) - (R + St_0) e^{-t/t_0}$

or $\quad T = a + b(t - t_0) + de^{-t/t_0}$

where $a = R$, $b = -S$, $d = -(R + St_0)$.

For a maximum or minimum

$$\frac{dT}{dt} = 0 = b - \frac{d}{t_0} e^{-t/t_0}$$

or

$$b = \frac{d}{t_0} e^{-t/t_0}.$$

You can show that $\dfrac{d^2 T}{dt^2}$ is *negative* for a maximum by substitution.

$$t = t_0 \ln \frac{d}{bt_0} = \frac{mc}{hA_S} \ln \left\{ \frac{-\left[\left(\frac{\dot{Q}_{g,0}}{hA_S} + T_f\right) + \frac{2.5 \frac{kW}{s} t_0}{hA_S}\right]}{-\left(\frac{2.5 \frac{kW}{s} t_0}{hA_S}\right)} \right\}$$

$$\left\{ t = \frac{mc}{hA_S} \ln \left(\frac{\frac{\dot{Q}_{g,0}}{hA_S} + T_f}{\frac{2.5 \frac{kW}{s} t_0}{hA_S}} + 1 \right) \right\}.$$

Now $\frac{\dot{Q}_{g,0}}{hA_S} + T_f = \frac{100 \text{ kW}}{0.0454 \frac{\text{kW}}{\text{m}^2 \text{ K}} \times 0.743 \text{ m}^2} + 288 \text{ K} = 3252.5 \text{ K}$

and $\frac{2.5 \frac{kW}{s} t_0}{hA_S} = \frac{2.5 \frac{kW}{s} \frac{mc}{hA_S}}{hA_S} = \frac{2.5 \frac{\text{kW}}{\text{s}} \left(\frac{22.7 \text{ kg} \times 0.377 \frac{\text{kJ}}{\text{kg K}}}{0.0454 \frac{\text{kW}}{\text{m}^2 \text{ K}} \times 0.743 \text{ m}^2} \right)}{0.0454 \frac{\text{kW}}{\text{m}^2 \text{ K}} \times 0.743 \text{ m}^2} \left[\frac{\text{kW s}}{\text{kJ}}\right]$

i.e. $\frac{2.5 \frac{kW}{s} t}{hA_S} = 18\,802.6 \text{ K}.$

Thus $t = \frac{22.7 \text{ kg} \times 0.377 \frac{\text{kJ}}{\text{kg K}}}{0.0454 \frac{\text{kW}}{\text{m}^2 \text{ K}} \times 0.743 \text{ m}^2} \ln \left(\frac{3252.5}{18\,802.6} + 1 \right)$

or $t = 253.7 \text{ s} \times 0.1595 = 40.5 \text{ s}.$

12.15 Some Properties of Metal and Alloys

Data for the preceding questions on heat transfer has, in the main, been extracted from the accompanying table of properties.

	k/[kW/(m K)]			c/[kJ/(kg K)]	ρ/(kg/m^3)
Material	0 °C	100 °C	300 °C	0 °C	0 °C
Metals					
Aluminium	0.2025	0.206	0.2302	0.871	2 707
Bismuth	0.0085	0.0068		0.121	9 803
Copper	0.3877	0.3774	0.367	0.381	8 938
Gold	0.2925	0.2943		0.126	19 270
Iron	0.062	0.0634		0.435	7 865
Lead	0.0348	0.0329	0.0312	0.126	11 293
Magnesium	0.1575	0.1593		0.971	1 746
Mercury	0.083			0.138	13 600
Nickel	0.0597	0.0589	0.0554	0.431	8 890
Silver	0.4189	0.412		0.234	10 492
Tin	0.0623	0.0589		0.226	7 304
Zinc	0.1125	0.1108	0.1021	0.381	7 144
Alloys					
Admiralty metal	0.1125	0.1108			
Brass (70/30)	0.0969	0.1039	0.1142	0.385	8 522
Bronze (75/25)	0.026			0.343	8 650
Cast iron					
plain	0.0571	0.055	0.0479	0.461	7 592
alloy	0.0519	0.049	0.0467	0.419	7 288
Constantan (60/40)	0.0215	0.0222		0.419	8 922
Steel					
type 18/8	0.0138	0.0163	0.0189	0.461	7817
type 347	0.0138	0.0161	0.0190	0.461	7817
mild 1%	0.0459	0.045	0.0433	0.461	7849

12.16 Dimensional Analysis in Heat Transfer

As indicated earlier in this chapter, Newton's law of cooling, namely

$$\dot{Q}'' = h \,.\, \Delta T, \qquad (h = \text{local heat transfer coefficient})$$

is deceptively simple.

h is a function of many variables and its value and the variables concerned are different for differing heat transfer configurations. There is a great difference between, for example, natural convection heat transfer past a vertical flat plate and forced convection over a bank of tubes.

Quite clearly, the real emphasis in the above law should be changed to

$$h = \frac{\dot{Q}''}{\Delta T}.$$

Thus the first problem is to find and tabulate values of h for as many configurations as possible which are of importance to engineers. A very large library of work is now in existence which testifies to the many experiments already carried out.

In order, however, to make the handling of the information concerning h more easily manageable, the engineer makes use of dimensional reasoning and dimensionless groups. The arguments which give rise to these groups may be found in the texts and will hopefully be clarified by some discrete examples in the following pages.

It is quite simply a technique, based upon dimensional reasoning, to enable engineers to present a large body of information in a compact form for easy reference.

12.17 Worked Examples

Example 12.9

Show that in natural convection heat transfer

$$Nu = \emptyset(Gr, Pr)$$

using variables μ, ρ, k, c_p, θ, βg, L where β is the coefficient of cubical expansion,

$$\beta = \frac{1}{\mathrm{d}T}\,\frac{\mathrm{d}v}{v},$$

and g is the gravitational acceleration.

Answer

The first and vital step is of course to specify correctly all the relevant variables. Assuming the list given is correct and complete, we may write generally

$$h = \emptyset(\mu, \rho, k, c_p, \theta, \beta g, L)$$

and reference to the texts on dimensional analysis will show that the next stage is to rewrite

$$h \propto \mu^{a} \,.\, \rho^{b} \,.\, k^{c} \,.\, c_p^{d} \,.\, \theta^{e} \,.\, (\beta g)^{f} \,.\, L^{g}.$$

Substituting dimensions Q (for heat), T (for temperature), and M, L, t for mass, length and time, we can write

$$\left(\frac{Q}{tL^2 T}\right)^{1} = \left(\frac{M}{Lt}\right)^{a} \cdot \left(\frac{M}{L^3}\right)^{b} \cdot \left(\frac{Q}{LtT}\right)^{c} \cdot \left(\frac{Q}{MT}\right)^{d} \cdot \left(T\right)^{e} \cdot \left(\frac{L}{Tt^2}\right)^{f} \cdot \left(L\right)^{g}.$$

$\left(\text{Note } \beta \text{ has dimensions } \dfrac{1}{T} \text{ and } \beta g \left(\dfrac{L}{Tt^2}\right)\right).$

Balancing the dimensions on both sides we now have

Heat (Q)	$1 = c + d$.	
Time (t)	$-1 = -a - c - 2f$	(or $1 = a + c + 2f$).
Length (L)	$-2 = -a - 3b - c + f + g$	(or $2 = a + 3b + c - f - g$).
Temperature (T)	$-1 = -c - d + e - f$	(or $1 = c + d - e + f$).
Mass (M)	$0 = a + b - d$.	

In the above we have five equations for seven unknowns so that a solution will be possible only in terms of groups of dimensions. The object is to arrange the solution to produce dimensionless groups where possible in order that dimensional similarity techniques can subsequently be used (e.g. as in model testing).

Thus we can now solve as follows:

c = 1 − d;
a = 1 − c − 2f = 1 − 1 + d − 2f = d − 2f;
b = d − a = d − d + 2f = 2f;
e = c + d + f − 1 = 1 − d + d + f − 1 = f;
g = a + 3b + c − f − 2 = d − 2f + 6f + 1 − d − f − 2 = −1 + 3f;

Note: The choice of which variables to select for retention (in this case d and f) is important but not necessarily exclusive. In this regard only practice makes proficient.

We can now write

$$h \propto \mu^{d-2f} \, . \, \rho^{2f} \, . \, k^{1-d} \, . \, c_p^{d} \, . \, \theta^{f} \, . \, (\beta g)^{f} \, . \, L^{3f-1}$$

or, rearranging in dimensionless groups,

$$\left(\frac{hL}{k}\right)^{1} \propto \left(\frac{\mu c_p}{k}\right)^{d} \cdot \left(\frac{L\,(\beta g)\,\theta\,\rho^2}{\mu^2}\right)^{f}$$

i.e. $$Nu = \emptyset\,(Pr), (Gr),$$

where $Nu \equiv$ Nusselt number $\left(\frac{hL}{k}\right)$, $Pr \equiv$ Prandtl number $\left(\frac{\mu c_p}{k}\right)$, $Gr \equiv$ Grashof number $\left(\frac{L\,(\beta g)\,\theta\,\rho^2}{\mu^2}\right)$.

Notes

(a) If irrelevant variables had originally been specified in the above analysis they would have yielded a zero power and subsequently have disappeared.

(b) L would be some characteristic linear dimension of the system (e.g. length of a vertical plate, diameter of a cylinder, et.).

(c) βg arises because of the buoyancy forces existing in natural correction owing to changes in density.

Example 12.10

Show that the dimensionless heat transfer coefficient for laminar flow through a circular tube is given by:

$$Nu = \emptyset\left(Pe, \frac{d}{L}\right)$$

where Pe = Peclet number = $\frac{ud}{\alpha}$, Nusselt (Nu) = $\frac{hd}{k}$, and $\alpha = \frac{k}{\rho C}$.

d is the inside tube diameter, L the tube length, and other symbols have their usual meaning.

Use only dimensions M, L, t and T.

When the inside wall temperature is close to the bulk fluid temperature then the equation is

$$Nu = 1.86\left(Pe\,.\,\frac{d}{L}\right)^{1/3}.$$

If the velocity of liquid ammonia through a tube of bore 0.3 cm and length 0.6 m is 7.5 cm/s, calculate the velocity of flow of water in the same tube for the same value of h.

Data:

Liquid	ρ/(kg/m³)	c/[kJ/(kg K)]	k/[kW/(m K)]
Ammonia	480.5	5.652	0.381×10^{-2}
Water	963.2	4.212	0.680×10^{-2}

Answer

Let $$h = \emptyset\,(d, k, u, \rho, c, L)$$

i.e. $$h \propto d^a \,.\, k^b \,.\, u^c \,.\, \rho^d \,.\, c^e \,.\, L^f$$

i.e. $$\left(\frac{M}{t^3 T}\right) \propto (L)^a \,.\, \left(\frac{ML}{t^3 T}\right)^b \,.\, \left(\frac{L}{t}\right)^c \,.\, \left(\frac{M}{L^3}\right)^d \,.\, \left(\frac{L^2}{t^2 T}\right)^e \,.\, (L)^f.$$

Length (L) $0 = a + b + c - 3d + 2e + f.$
Mass (M) $1 = b + d.$
Time (t) $-3 = -3b - c - 2e$ or $3 = 3b + c + 2e.$
Temperature (T) $-1 = -b - e$ or $1 = b + e.$

$b = 1 - e;$
$d = 1 - b = 1 - 1 + e = e;$
$c = 3 - 3(1 - e) - 2e = e;$
$a = -b - c + 3d - 2e - f = e - 1 - e + 3e - 2e - f = -1 + e - f.$

Thus $$h \propto (d^{-1+e-f}) \,.\, (k)^{1-e} \,.\, (u)^e \,.\, (\rho)^e \,.\, (c)^e \,.\, (L)^f$$

or $$\frac{hd}{k} \propto \left(\frac{d\,\rho\,u\,c}{k}\right)^e \,.\, \left(\frac{L}{d}\right)^f$$

or $$Nu = \emptyset\left(\frac{\rho c}{k} \times ud\right), \left(\frac{L}{d}\right) = \emptyset\left(\frac{ud}{\alpha}\right), \left(\frac{d}{L}\right).$$

Note that $\frac{L}{d}$ can be inverted to $\frac{d}{L}$ in this kind of expression since the function is perfectly general.

In the given case

$$Nu = 1.86\left(Pe \,.\, \frac{d}{L}\right)^{1/3}$$

and if a refers to ammonia and w refers to water then:

$$\frac{h_a d_a}{k_a} = 1.86\left(Pe_a \,.\, \frac{d}{L}\right)^{1/3}; \qquad \frac{h_w d_w}{k_w} = 1.86\left(Pe \,.\, \frac{d}{L}\right)^{1/3}.$$

Given that $$h_a = h_w$$

and for dimensional similarity $$Pe_a = Pe_w$$

then $$\frac{k_w}{k_a} = \left(\frac{Pe_a}{Pe_w}\right)^{1/3}$$

or $$\left(\frac{0.68}{0.381}\right)^3 = \frac{u_a\, d_a\, \rho_a\, c_a\, k_w}{u_w\, d_w\, \rho_w\, c_w\, k_a} \quad \text{and} \quad d_a = d_w$$

or $$\left(\frac{0.68}{0.381}\right)^3 = \frac{u_a}{u_w} \times \frac{480.5}{963.2} \times \frac{5.652}{4.212} \times \frac{0.68}{0.381}$$

or $$u_w = 7.5\ \frac{\text{cm}}{\text{s}}\ (0.499)\,(1.342)\,(0.314) = 1.577\ \frac{\text{cm}}{\text{s}}.$$

Example 12.11

For forced-convection heat transfer problems the heat transfer coefficient h is found to depend only on the fluid viscosity μ and fluid density ρ, the thermal conductivity of the fluid k, the specific heat capacity of the fluid c, a characteristic dimension of the solid surface l, the temperature difference between the surface and the fluid θ, and the fluid velocity u.

Show, by dimensional reasoning that one form of the relationship may be expressed:

$$h = \frac{k}{l}\,\emptyset\left(\frac{\mu c}{k}, \frac{\rho u l}{\mu}, \frac{u^2}{c\theta}\right) \qquad \text{or} \qquad Nu = \emptyset\,(Pr, Re, Ec)$$

where Nu = Nusselt number = hl/k;
Pr = Prandtl number = $\mu c/k$;
Re = Reynolds number = $\rho u l/\mu$;
Ec = Eckert number = u^2/c.

The heat transfer coefficient for a forced-convection air cooler system is to be estimated from a test on a one-third scale model using air. If the velocity of the air in the cooler is to be 12 m/s, calculate the corresponding air velocity in the model test.

If, in the model test, the heat transfer rate is measured as 33.33 kW when the cooling surface area is 4 m^2 and the temperature difference between the air and the surface is 25 K, calculate the heat transfer coefficient for the prototype cooler. Assume that compressibility effects are negligible and that fluid properties are the same for both model and prototype.

Answer

$$h \propto \mu^{a_1} . \rho^{a_2} . k^{a_3} . c^{a_4} . l^{a_5} . \theta^{a_6} . u^{a_7} .$$

Mass (M) $\quad 1 = a_1 + a_2 + a_3$.
Length (L) $\quad 0 = -a_1 - 3a_2 + a_3 + 2a_4 + a_5 + a_7$.
Time (t) $\quad -3 = -a_1 - 3a_3 - 2a_4 - a_7$.
Temperature (T) $\quad -1 = -a_3 - a_4 + a_6$.

$a_4 = 1 - a_3 + a_6$;
$a_1 = 3 - 3a_3 - 2(1 - a_3 - a_6) - a_7 = 1 - a_3 + 2a_6 - a_7$;
$a_2 = 1 - a_1 - a_3 = 1 - (1 - a_3 + 2a_5 - a_7) - a_3 = -2a_6 + a_7$;
$a_5 = a_1 + 3a_2 - a_3 - 2a_4 - a_7 = 1 - a_3 + 2a_6 - a_7 + 3(-2a_6 + a_7) - a_3$
$\quad - 2(1 - a_3 - a_6) - a_7$;
$a_5 = -1 + a_7 - 2a_6$.

Thus $$h = \mu^{1-a_3+2a_6-a_7} . \rho^{-2a_6+a_7} . k^{a_3} . c^{1-a_3+a_6} . L^{-1+a_7-2a_6} . \theta^{a_6} . u^{a_7}$$

or $$h = \frac{k}{L}\mu^{1-a_3+2a_6-a_7}\, \rho^{-2a_6+a_7} . k^{a_3-1} . c^{1-a_3+a_6} . L^{a_7-2a_6} . \theta^{a_6} . u^{a_7}$$

or $$h = \frac{k}{L}\left[\left(\frac{\mu c}{k}\right)^{1-a_3} \times \left(\frac{\rho u L}{\mu}\right)^{a_7-2a_6} \times \left(\frac{c\theta}{u^2}\right)^{a_6}\right]$$

or $$h = \frac{k}{L}\,\emptyset\left[\frac{\mu c}{k}, \frac{\rho u L}{\mu}, \frac{u^2}{c\theta}\right] = \frac{k}{L}\,\emptyset\,(Pr, Re, Ec).$$

There can be no change in fluid properties since there are no changes in Pr, Re or Ec numbers for dimensional similarity between the model and prototype.

i.e. $\dfrac{h_{\text{prototype}}}{h_{\text{scale model}}} = \dfrac{l_{\text{scale model}}}{l_{\text{prototype}}} = \dfrac{1}{3}$

or $h_{\text{model}} = 3h_{\text{prototype}}$

and $u_{\text{model}} = 3u_{\text{prototype}} = 36\ \dfrac{\text{m}}{\text{s}}$ $\left(\text{since} \left(\dfrac{\rho u L}{\mu}\right)_{\text{model}} = \left(\dfrac{\rho u L}{\mu}\right)_{\text{prototype}}\right)$.

Also $h_{\text{scale model}} = \dfrac{33.33\ \text{kW}}{4\text{m}^2\ 25\ \text{K}} = 0.1111\ \dfrac{\text{kW}}{\text{m}^2\ \text{K}} = 111.1\ \dfrac{\text{W}}{\text{m}^2\ \text{K}}$.

Exercises

1 The interior of a coldroom is maintained at $-40\ °\text{C}$, the walls being of brick, insulated internally with fibreglass which is lined with aluminium.

Calculate, for an outside temperature of $20\ °\text{C}$,

(a) the heat transfer rate, per unit area of wall, into the room and
(b) the temperature of the brick and fibreglass interface.

Data:		Thickness/mm	Conductivity/[W/(m K)]
	Brick	114	1.16
	Fibreglass	50	0.04
	Aluminium	2.5	199.0

Surface heat transfer coefficients:
Aluminium to inside cold air 6.8 W/(m² K)
Outside air to brick 22.7 W/(m² K)

(39 W/m, 14.5 °C)

2 Show that for radial steady conduction through a hollow sphere of inside radius r_1, and outside radius r_2, and corresponding temperatures at these radii of T_1 and T_2, the rate of heat transfer is given by

$$\dot{Q} = \frac{4\pi r_1\ r_2\ k\ (T_2 - T_1)}{r_2 - r_1}$$

where k = thermal conductivity.

A sphere is lagged from a radius of 0.4 m to a radius of 0.5 m with inner and outer surface temperatures of 232 °C and 65 °C respectively. If k for the lagging is 0.066 W/(m K) calculate the rate of heat transfer. (277 W)

3 A large nickel plate 2.5 cm thick is heated to a uniform temperature of 1100 °C prior to cold working the metal in ambient air at 26 °C. Neglecting the internal thermal resistance of the plate and taking a uniform h of 0.034 kW/(m² K) show that the working process must be completed within approximately 4.3 minutes if the minimum cold working temperature of nickel is 920 °C.

4 It is expedient during a test to use the starter of an electric motor as a speed control. The power dissipated in the lacquer-insulated coils is 200 W under the test conditions. If the initial temperature of the coils is 15 °C and the lacquer begins to melt at 200 °C, find the maximum duration of the test and the further time needed for the coils to cool to 40 °C.

Data: Mass of coils 0.5 kg
Specific heat of coils 0.419 kJ/(kg K)
Surface area of coils 0.0025 m^2
Surface h 0.034 kW/(m^2 K)
Air temperature 15 °C

(3.36 min, 82.2 min)

5 An electrically powered relay operates repetitively. The cycle times are 1 second with power ON followed by 9 seconds with power OFF. The thermal time constant of the relay is known to be 30 seconds and if the power is left on continuously then the steady-state temperature of the coils above the temperature of the surrounding air is 133 °C.

Estimate the maximum and minimum temperatures which recur during continuous normal operation if the air temperature is 21 °C. (36.4 °C, 32.4 °C)

6 Show that the transient temperature distribution in a long cylinder may be expressed as:

$$\frac{hR}{k} = \phi\left(\frac{\alpha t}{R^2}, \frac{r}{R}, \frac{T_r}{T}\right)$$

where α = thermal diffusivity $\left(= \dfrac{k}{\rho c}\right)$,

$\dfrac{\alpha t}{R^2}$ = Fourier modulus,

$\dfrac{hR}{k}$ = Biot modulus (a form of Nusselt number),

r = radial coordinate,
R = outside radius,
t = time,
T_r = temperature at radius r,
T = initial temperature,

and h and k have the usual meanings.

13 Specimen Examination Paper Number 1

The following eight questions, based on the range of work covered in this volume, are given as they might appear in a typical examination paper at the end of a first year's work.

The solution using the problem solution approach, where appropriate, is given question by question immediately afterwards.

It is hoped that, like the rest of this book, you will always attempt to solve the problem for yourself first and only refer to the given solutions when in trouble.

13.1 Questions

1 (a) Complete the following table of properties for 'water substance' in its various phases:

$\frac{p}{\text{bar}}$	$\frac{T}{°\text{C}}$	$\frac{v}{\text{m}^3/\text{kg}}$	$\frac{e}{\text{kJ/kg}}$	$\frac{h}{\text{kJ/kg}}$	Regime or quality
	300		2774		
10		0.001 07			
	195			2000	

(b) Superheated mercury vapour may be considered to behave like a perfect gas. Calculate the specific enthalpy, specific volume and specific internal energy at 30 bar, 700 °C.

2 0.5 kg of a perfect gas occupies 0.1 m³ at 120 °C and 370 kN/m². If the ratio of the two principal specific heat capacities is 1.25, determine the values of each. The gas is heated isochorically till its pressure is 1480 kN/m² and then expanded reversibly and adiabatically to the original pressure after which it is cooled to the original volume. Sketch the cycle on a p–v diagram and calculate the net work transfer between the gas and surroundings.

3 Show that for purely radial conduction of heat in a right circular cylinder of thermal conductivity k

$$T_1 - T_2 = \frac{\dot{Q}}{2\pi Lk} \ln \frac{r_2}{r_1}$$

where T is the temperature at radius,
and $\dot{Q}$ is the rate of heat transmission.

The power supplied to an electric cooker is carried by copper conductors of 2.3 mm diameter covered by a concentric layer of polythene insulation and an outer shield of PVC. The thickness of polythene is 0.5 mm and its thermal conductivity is 0.35 W/(m K); the PVC casing is 1 mm thick with a thermal conductivity of 0.20 W/(m K). I^2R losses in the copper are conducted through the plastics and convected from the outer surface of the PVC to the surrounding air which is at 40°C, the surface heat transfer coefficient being 28 W/(m^2 K).

Determine the temperature of the inner surface of PVC when the current carried is 60 amperes. The electrical resistance of the copper wire is 3.055×10^{-3} Ω/m.

4 The percentage volume analysis of a fuel gas is as follows:

H_2, 40; CH_4, 27; CO, 13; CO_2, 3; O_2, 5; N_2, 12.

Find the stoichiometric air : fuel ratio by volume.

The gas is now mixed with 9 times its volume of air to form a mixture for an internal-combustion engine.

Assuming complete combustion determine:

(a) the percentage excess air supplied (by volume),
(b) the volumetric analysis of the wet postcombustion gases,
(c) the partial pressures of the constituents of the postcombustion mixture at 1.1 bar and 77 °C and the dewpoint temperature.

5 Steam flows through a nozzle which turns the flow through an angle of 90° as in the figure. At entry the steam is at 50 bar and 500 °C, the cross-sectional area is 15 cm^2 and the velocity of flow is 20 m/s. At exit the pressure is 0.5 bar, the dryness 0.98 and the cross-sectional area is 10.7 cm^2.

Determine the net force exerted on the bolted joints at either end of the nozzle by the flowing steam.

Assume atmospheric pressure outside the nozzle.

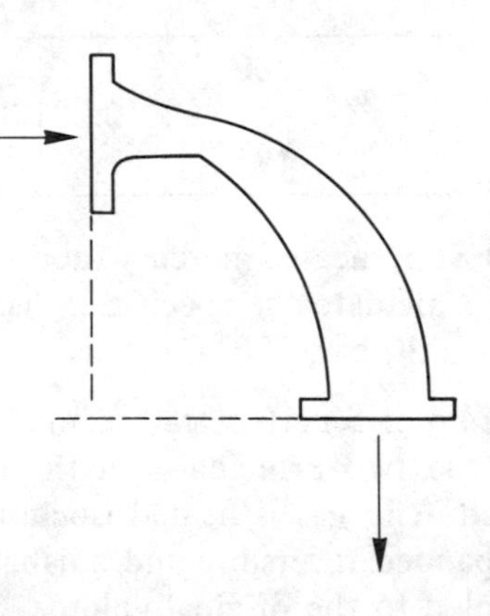

6 A perfectly insulated receiver is supplied with two, separate, steady streams of air A and B, and a single, steady stream of air C leaves it with a low velocity. There is a work transfer into the receiver at the rate of 500 kW.

Calculate the temperature of the delivery in degrees C, assuming air is a perfect gas and given the following data:

Stream A: Supplied through a circular pipe 0.3 m diameter at 1 bar, 50 °C at 15 m/s.

Stream B: Supplied through a circular pipe 0.4 m diameter at 1.5 bar, 100 °C at 20 m/s.

7 Steam contained in a cylinder initially occupies a volume of 0.5 m^3 at a pressure of 1.5 bar and temperature 165 °C. If the steam then undergoes a reversible isothermal compression process until its final volume is one sixth of the initial volume, calculate:

(a) the change in internal energy,
(b) the change in entropy,
(c) the heat transfer.

8 A fine tungsten wire, 5 μm in diameter by 3 mm long, is heated electrically and the energy dissipated is stored and/or convected to air flowing normal to the axis of the wire. If the convective heat transfer coefficient from wire to air is 4000 W/(m^2 K) and the density of tungsten is 19 000 kg/m^3 calculate the thermal time constant of the wire. c for the wire is 340 J/(kg K).

The d.c. electrical supply is switched 'on' and 'off' for equal periods to give a square-wave input. Sketch the expected form of the variation of wire temperature with time. If the wire temperature is to be 300 °C ± 1 °C and the air temperature is 20 °C determine:

(a) the 'forcing temperature' during the 'on' period,
(b) the frequency of the switching cycle,
(c) the necessary supply voltage if the resistivity of tungsten is 50×10^{-8} Ω m.

13.2 Suggested Solutions

Question 1

This is an example on property values.

(a) Line 1: refer to tables page 5.

At 300 °C, $$2565 \frac{kJ}{kg} \leqslant e_g \leqslant 2559 \frac{kJ}{kg}.$$

But $$e = 2774 \frac{kJ}{kg}.$$

Thus we have superheated steam and reference to p. 7 gives

$$p = 20 \text{ bar}; \qquad v = 0.1255 \frac{m^3}{kg}; \qquad h = 3025 \frac{kJ}{kg}.$$

Line 2: at p = 10 bar, $v_g = 0.1944 \frac{m^3}{kg}$. (p. 4)

at p = 10.03 bar, $v_f = 0.0011\,28 \frac{m^3}{kg}$. (p. 10)

But $$v = 0.001\,07 \frac{m^3}{kg} \quad (< v_f).$$

Thus we have a subcooled liquid and T is the important variable. Assume we hav a saturated liquid at 130 °C with given value of v as tabulated (since p has negligible effect on v). Now, referring to page 4 at T = 130 °C, we read

$$\left.\begin{aligned} e_f &= 546 \frac{kJ}{kg}, \\ h_f &= 546 \frac{kJ}{kg}. \end{aligned}\right\}$$

Line 3: at $T = 195\,°\text{C}$ (p. 4), $h_g = 2790$ kJ/kg; $h_f = 830$ kJ/kg.

But $$h = 2000\ \frac{\text{kJ}}{\text{kg}}, \rightarrow \text{thus wet vapour}$$

$$x = \frac{h - h_f}{h_{fg}} = \frac{2000 - 830}{1960} = 0.597,$$

$$p = 14 \text{ bar.}$$

$$v \triangleq xv_g \quad \text{(ignoring } v_f) \quad \triangleq 0.597 \times 0.1408 = 0.0841\ \frac{\text{m}^3}{\text{kg}}.$$

$$e = h - pv = 2000\ \frac{\text{kJ}}{\text{kg}} - \left(1400\ \frac{\text{kN}}{\text{m}^2} \times 0.0841\ \frac{\text{m}^3}{\text{kg}}\right)\left[\frac{\text{kJ}}{\text{kN m}}\right] = 1882.3\ \frac{\text{kJ}}{\text{kg}}.$$

(b) *Gas:* use page 14 for data for mercury.

And at 30 bar $h_g = 371.36\ \frac{\text{kJ}}{\text{kg}}$; $c_p = 0.1036\ \frac{\text{kJ}}{\text{kg K}}$ (foot of p. 14).

Thus $$h = h_g + c_p(T - T_{sat}) = 371.36\ \frac{\text{kJ}}{\text{kg}} \times 0.1036\ \frac{\text{kJ}}{\text{kg K}}\ (700 - 630)\ \text{K}$$

$$= 378.61\ \frac{\text{kJ}}{\text{kg}}.$$

Now $$v = v_g \times \frac{T}{T_{sat}} = 0.01252\ \frac{\text{m}^3}{\text{kg}} \times \left(\frac{700 + 273}{630 + 273}\right) = 0.0135\ \frac{\text{m}^3}{\text{kg}}$$

and $$e = h - pv = 378.61\ \frac{\text{kJ}}{\text{kg}} - \left(3000\ \frac{\text{kN}}{\text{m}^2} \times 0.0135\ \frac{\text{m}^3}{\text{kg}}\right)\left[\frac{\text{kJ}}{\text{kN m}}\right]$$

$$= 338.11\ \frac{\text{kJ}}{\text{kg}}.$$

Question 2

There are three *non-flow processes* constituting a cycle here.
The fluid is a *perfect gas.*
The most useful *picture* is probably a sketch of the *p/v field.*

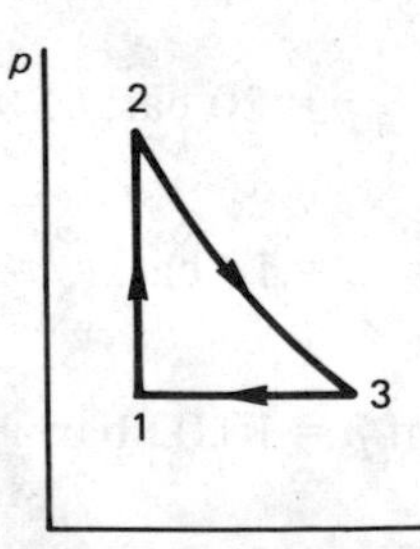

The first part of the problem deals with property calculation.

$$\text{for a gas,} \quad R = \frac{p_1 V_1}{m T_1} = \frac{370\ \frac{\text{kN}}{\text{m}^2} \times 0.1\ \text{m}^3}{0.5\ \text{kg} \times 393\ \text{K}} = 0.1883\ \frac{\text{kJ}}{\text{kg K}} = c_p - c_v.$$

But $\frac{c_p}{c_v} = 1.25 = \gamma$ and substituting from one result into the other we get

$$c_v = 0.7532\ \text{kJ/(kg K)},$$

$$c_p = 0.9415\ \text{kJ/(kg K)}.$$

Energy equation: $W_{net} = {}_1Q_2 + {}_3Q_1$ (since ${}_2Q_3 = 0$ adiabatic).

$${}_1Q_2 = m(e_2 - e_1) = m\,c_v\,(T_2 - T_2) \qquad [{}_1W_2 = 0].$$

Now $\dfrac{T_2}{T_1} = \dfrac{p_2}{p_1} = \dfrac{1480}{370} = 4$ $(V_1 = V_2)$,

$$T_2 = 4 \times 393\ \text{K} = 1572\ \text{K};$$

$${}_1Q_2 = 0.5\ \text{kg} \times 0.7532\ \frac{\text{kJ}}{\text{kg K}}\ (1572 - 393)\ \text{K} = 444\ \text{kJ}.$$

Also, $T_3 = T_2 \left(\dfrac{p_3}{p_2}\right)^{(\gamma-1)/\gamma} = 1572\ \text{K} \left(\dfrac{370}{1480}\right)^{0.2} = 1191.4\ \text{K}.$

Thus ${}_3Q_1 = mc_p\,(T_1 - T_3)$ (at constant pressure)

$${}_3Q_1 = 0.5\ \text{kg} \times 0.9415\ \frac{\text{kJ}}{\text{kg K}}\ (393 - 1191.4)\ \text{K} = -\ 375.8\ \text{kJ}$$

and $W_{net} = 444 - 375.8 = +\ 68.2$ kJ.

Question 3

This is thermal conduction in the steady state (i.e. *not* thermodynamics).

A picture showing the temperature distribution will help.

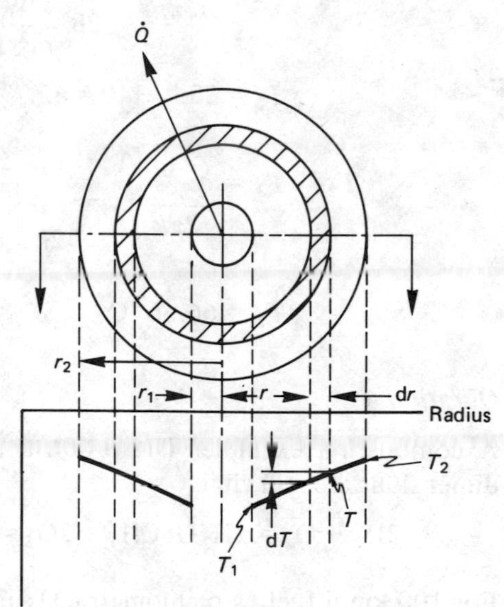

Fourier:

$$\dot{Q} = -\,kA\,\frac{dT}{dr} = -\,k2\pi rL\,\frac{dT}{dr}$$

or

$$\int_{r_1}^{r_2} \dot{Q}\,\frac{dr}{r} = \int_{T_1}^{T_2} -\,k2\pi L\,dT$$

or

$$\dot{Q}\ln\frac{r_2}{r_1} = -\,k2\pi L\,(T_2 - T_1)$$

or

$$T_1 - T_2 = \frac{\dot{Q}\ln(r_2/r_1)}{2\pi kL}.$$

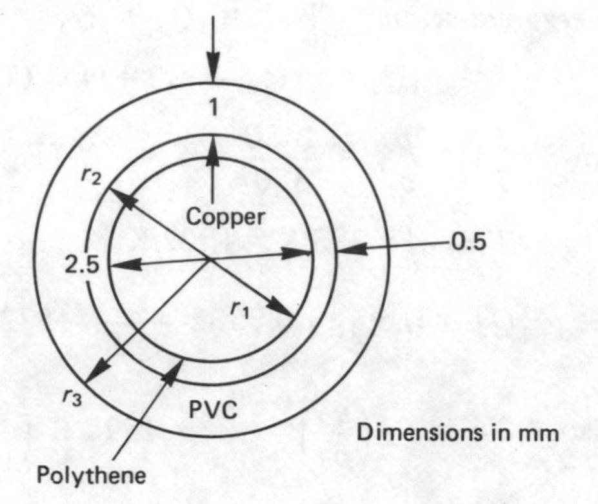

$$I^2R = 60^2 \text{ A}^2 \times 3.055 \times 10^{-3} \frac{\Omega}{\text{m}} \left[\frac{\text{V}}{\text{A}\,\Omega}\right] \left[\frac{\text{W}}{\text{A V}}\right]$$

$$I^2R = 11 \frac{\text{W}}{\text{m}} = \dot{Q}'.$$

$\dot{Q}' = h2\pi r_3 (T_3 - T_a)$ from surface of PVC to atmosphere

or
$$T_3 - T_a = \frac{\dot{Q}'}{h2\pi r_3} = \frac{11 \frac{\text{W}}{\text{m}}}{28 \frac{\text{W}}{\text{m}^2\text{ K}} 2\pi \times 0.002\,75 \text{ m}} = 22.7 \text{ K},$$

$$T_3 = 22.7 + 40 = 62.7\,°\text{C},$$

$$T_2 - T_3 = \frac{\dot{Q}' \ln \frac{r_3}{r_2}}{2\pi k_{23}} = \frac{11 \frac{\text{W}}{\text{m}} \ln \frac{2.75}{1.75}}{2\pi \times 0.2 \frac{\text{W}}{\text{m K}}} = 3.96 \text{ K},$$

$$T_2 = 66.66\,°\text{C}.$$

Question 4

A combustion example. The problem solution approach is of less use here but dimensions are still vital.

$$2H_2 + O_2 = 2H_2O; CH_4 + 2O_2 = CO_2 + 2H_2O; 2CO + O_2 = 2CO_2.$$

For 100 kmol fuel, stoichiometric O_2 needed = $\frac{40}{2} + (27 \times 2) + \frac{13}{2} - 5.$

(for H_2; for CH_4; for CO; in fuel as O_2)

Stoich. O_2 = 75.5 kmol;

stoich. $\frac{n_{air}}{n_{fuel}} = \frac{75.5}{100} \frac{\text{kmol } O_2}{\text{kmol fuel}} \left\{\frac{\text{kmol air}}{0.21 \text{ kmol } O_2}\right\}$ ← (p. 24 of tables)

Stoich. $\frac{n_{air}}{n_{fuel}} = 3.595 \frac{\text{kmol } O_2}{\text{kmol fuel}}.$

Actual air supply = 9 kmol air/kmol fuel.
Excess air supply = 9 − 3.595 = 5.405 kmol air/kmol fuel

$$\text{Percentage excess} = \frac{5.405}{3.595} \times 100 = 150.3 \text{ per cent by volume.}$$

The postcombustion gases (for 100 kmol fuel) will be

$$(40 \text{ kmol } H_2O)_{\text{from } H_2} + (54 \text{ kmol } H_2O)_{\text{from } CH_4} = 94 \text{ kmol } H_2O$$
$$+ (27 \text{ kmol } CO_2)_{\text{from } CH_4} + (13 \text{ kmol } CO_2)_{\text{from } CO} + (3 \text{ kmol } CO_2)_{\text{in fuel}} = 43 \text{ kmol } CO_2$$
$$+ (0.21 \times 540.5) \text{ kmol } O_2 \text{ excess}$$
$$+ (0.79 \times 900) \text{ kmol } N_2 \text{ in total air supply.}$$

Thus $\eta_{total} = 961.5$ kmol of wet combustion gases.

$$\text{Thus } x(H_2O) = \frac{n(H_2O)}{n_{total}} \times 100 = \frac{94}{961.5} \times 100 = 9.78 \text{ per cent.}$$

$$\text{Also } x(CO_2) = \frac{4300}{961.5} = 4.47 \text{ per cent; } x(O_2) = \frac{0.21 \times 540.5}{961.5} \times 100 = 11.8 \text{ per cent}$$

$$x(N_2) = \frac{0.79 \times 900}{961.5} \times 100 = 73.9 \text{ per cent.}$$

$$p(H_2O) = x(H_2O) \times p_{total} = 0.0978 \times 1.1 = 0.1076 \text{ bar;}$$
$$p(CO_2) = x(CO_2) \times p_{total} = 0.0447 \times 1.1 = 0.0492 \text{ bar;}$$
$$p(O_2) = x(O_2) \times p_{total} = 0.118 \times 1.1 = 0.1298 \text{ bar;}$$
$$p(N_2) = 1.1 - (0.1076 + 0.0492 + 0.1298) = 0.8134 \text{ bar.}$$

The dew point at 0.1076 bar = T_{sat} = 47.2 °C. (Interpolating p. 3)

Question 5

The question is about forces on bolts on the nozzle. Clearly the momentum equation will be needed in the solution.

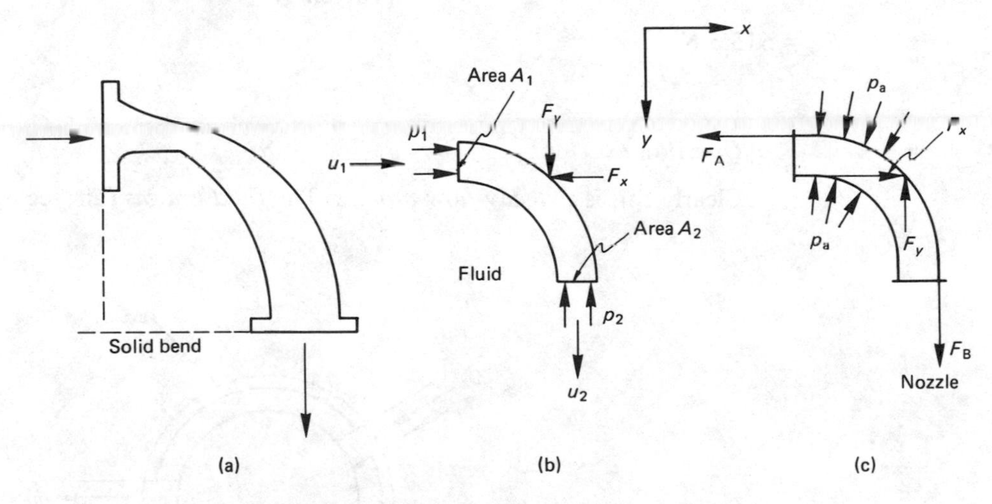

(a) (b) (c)

Applying *momentum* to the fluid 'plug' within the nozzle,

$$p_1 A_1 - F_x = \dot{m}(0 - u_1)$$

or $$F_x = p_1 A_1 + \dot{m} u_1. \quad \ldots (1)$$

Also, $$F_y - p_2 A_2 = \dot{m}(u_2 - 0)$$

or $$F_y = p_2 A_2 + \dot{m} u_2. \quad \ldots (2)$$

The nozzle is in equilibrium, hence

$$F_x - F_A - p_a A_1 = 0 \qquad \ldots (3)$$

and $$F_B - F_y + p_a A_2 = 0. \qquad \ldots (4)$$

From (1) and (3) $$F_A = (p_1 - p_a)A_1 + \dot{m}u_1 \qquad \ldots (5)$$

and from (2) and (4) $$F_B = (p_2 - p_a)A_2 + \dot{m}u_2. \qquad \ldots (6)$$

At 50 bar, 500 °C; $v_1 = 0.0685 \dfrac{\text{m}^3}{\text{kg}}$. (p. 7 of tables)

Mass continuity $$\dot{m} = \frac{A_1 u_1}{v_1} = \frac{15\ \text{cm}^2 \times 20\ \dfrac{\text{m}}{\text{s}}}{0.0685\ \dfrac{\text{m}^3}{\text{kg}}} \left[\frac{\text{m}^2}{10^4\ \text{cm}^2}\right]$$

$$\dot{m} = 0.438\ \frac{\text{kg}}{\text{s}}.$$

At 0.5 bar, 0.98 dry, $v_2 \simeq 0.98 \times 3.239 = 3.174\ \dfrac{\text{m}^3}{\text{kg}}$.

Thus $$u_2 = \frac{\dot{m}v_2}{A_2} = \frac{0.438\ \dfrac{\text{kg}}{\text{s}} \times 3.174\ \dfrac{\text{m}^3}{\text{kg}}}{10.7\ \text{cm}^2} \left[\frac{10^4\ \text{cm}^2}{\text{m}^2}\right]$$

$$u_2 = 1299.2\ \frac{\text{m}}{\text{s}}.$$

From (5) $$F_A = (50 - 1) \times 100\ \frac{\text{kN}}{\text{m}^2} \times 15\ \text{cm}^2 \left[\frac{\text{m}^2}{10^4\ \text{cm}^2}\right] + 0.438\ \frac{\text{kg}}{\text{s}} \times 20\ \frac{\text{m}}{\text{s}} \left[\frac{\text{N s}^2}{\text{kg m}}\right]$$

$$F_A = 7.36\ \text{kN}.$$

From (6) $$F_B = (0.5 - 1) \times 100\ \frac{\text{kN}}{\text{m}^2} \times 10.7\ \text{cm}^2 \left[\frac{\text{m}^2}{10^4\ \text{cm}^2}\right] + 0.438\ \frac{\text{kg}}{\text{s}} \times 1299.2\ \frac{\text{m}}{\text{s}} \left[\frac{\text{N s}^2}{\text{kg m}}\right]$$

$$F_B = 515.5\ \text{N}.$$

Question 6

Clearly this is a *steady-flow process*. The *fluid is a gas* (air; see p. 24).

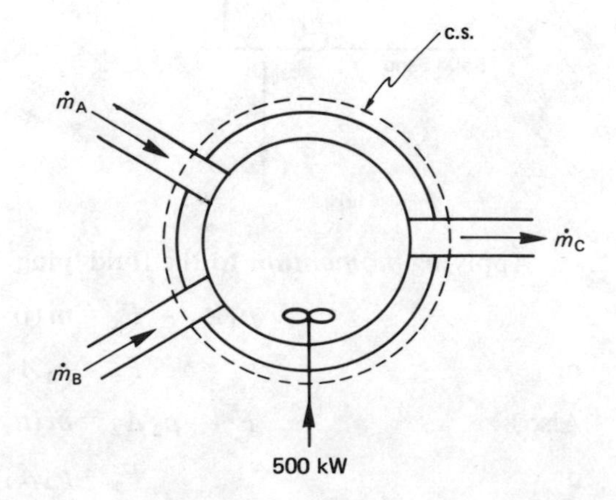

Assumptions:

(a) Although we need the values of u_A and u_B for the calculation of $\dot{m}_A$ and $\dot{m}_B$, they are small, cf. h_A and h_B.

(b) Δgz is negligible everywhere.

(c) $\dot{Q} = 0$.

Mass continuity: $\dot{m}_C = \dot{m}_A + \dot{m}_B = \dfrac{\dot{V}_A}{v_A} + \dfrac{\dot{V}_B}{v_B} = \left(\dfrac{Au}{v}\right)_A + \left(\dfrac{Au}{v}\right)_B$.

Steady-flow energy equation:

$$-\dot{W} = \dot{m}_C h_C - \dot{m}_A h_A - \dot{m}_B h_B \qquad \text{(all else zero).}$$

Gas equation:

$$v_A = \left(\frac{RT}{p}\right)_A = \frac{0.287 \dfrac{\text{kJ}}{\text{kg K}} \times 323 \text{ K}}{100 \dfrac{\text{kN}}{\text{m}^2}} = 0.927 \frac{\text{m}^3}{\text{kg}};$$

$$v_B = \left(\frac{RT}{p}\right)_B = \frac{0.287 \times 373}{150} = 0.714 \frac{\text{m}^3}{\text{kg}}.$$

$$\dot{m}_A = \left(\frac{\dot{V}}{v}\right)_A = \left(\frac{Au}{v}\right)_A = \frac{\pi \times 0.3^2 \text{ m}^2 \times 15 \dfrac{\text{m}}{\text{s}}}{4 \times 0.927 \dfrac{\text{m}^3}{\text{kg}}} = 1.144 \frac{\text{kg}}{\text{s}};$$

$$\dot{m}_B = \frac{\pi \times 0.4^2 \times 20}{4 \times 0.714} = 3.52 \text{ kg/s};$$

$$\dot{m}_C = 1.144 + 3.52 = 4.664 \text{ kg/s}.$$

$$\dot{m}_A h_A = \dot{m}_A c_p T_A = 1.144 \frac{\text{kg}}{\text{s}} \times 1.005 \frac{\text{kJ}}{\text{kg K}} \times 323 \text{ K} = 371.4 \text{ kW};$$

$$\dot{m}_B h_B = \dot{m}_B c_p T_B = 3.52 \times 1.005 \times 373 = 1319.5 \text{ kW}.$$

Now $\qquad \dot{W} = -500$ kW (work IN).

Thus $\qquad \dot{m}_C h_C = \dot{m}_C c_p T_C = 371.4 + 1319.5 + 500 = 2190.9$ kW

and

$$T_C = \frac{2190.9 \text{ kW}}{4.664 \dfrac{\text{kg}}{\text{s}} \times 1.005 \dfrac{\text{kJ}}{\text{kg K}}} = 467.4 \text{ K} = 194.4\,°\text{C}.$$

Question 7

This question is concerned with a *non-flow process* (steam in a closed cylinder).

The *fluid is steam.*

A useful picture might be the p–v diagram sketch.

First of all we need the mass: $m = \dfrac{V_1}{v_1}$;

$$v_1 = 1.286 + \frac{15}{50}(1.445 - 1.286) = 1.334 \frac{\text{m}^3}{\text{kg}}.$$

(Horizontal interpolation on p. 6 is acceptable.)

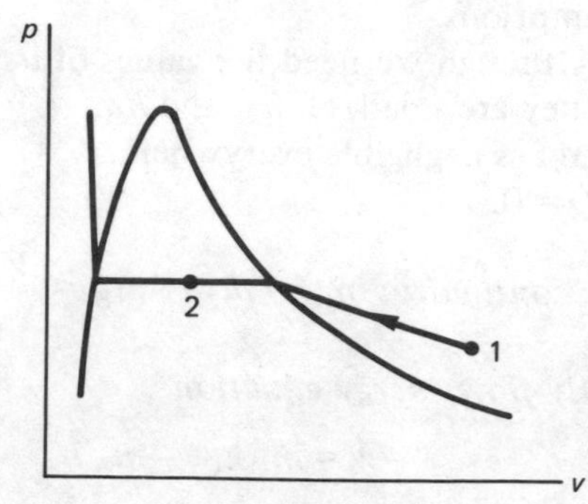

$$m = \frac{V_1}{v_1} = \frac{0.5 \text{ m}^3}{1.334 \dfrac{\text{m}^3}{\text{kg}}} = 0.375 \text{ kg}.$$

$$V_2 = \frac{V_1}{6} = 0.0833 \text{ m}^3.$$

$$v_2 = \frac{V_2}{m} = \frac{0.0833}{0.375} = 0.222 \frac{\text{m}^3}{\text{kg}} \; ; \; T_2 = 165 \text{ °C given.}$$

Thus $$x_2 \simeq \frac{v_2}{v_{g,2}} = \frac{0.222}{0.2728} = 0.813.$$

Now $$e_1 = 2580 + \frac{15}{50}(2656 - 2580) = 2602.8 \frac{\text{kJ}}{\text{kg}} \quad \text{(p. 6)}$$

and $$e_2 = 467 + 0.813(2519 - 467) = 2135.3 \frac{\text{kJ}}{\text{kg}} \quad \text{(p. 4)}$$

$$E_2 - E_1 = m(e_2 - e_1) = 0.375 \text{ kg}(2135.3 - 2602.8) \frac{\text{kJ}}{\text{kg}} = -175.3 \text{ kJ}. \quad \ldots (1)$$

$$s_1 = 7.42 + \frac{15}{50}(7.643 - 7.42) = 7.487 \frac{\text{kJ}}{\text{K kg}} \; ; \quad \text{(p. 6)}$$

$$s_2 = 1.434 + 0.813(5.789) = 6.140 \frac{\text{kJ}}{\text{K kg}}. \quad \text{(p. 4)}$$

$$S_2 - S_1 = m(s_2 - s_1) = 0.375 \text{ kg}(6.14 - 7.487) \frac{\text{kJ}}{\text{K kg}} = -0.505 \frac{\text{kJ}}{\text{K}}. \quad \ldots (2)$$

$${}_1Q_2 = m \int_1^2 T \, . \, \text{d}s = mT(s_2 - s_1) \quad \text{for an isothermal process}$$

$$= 0.375 \text{ kg} \times 438 \text{ K}(6.14 - 7.487) \frac{\text{kJ}}{\text{K kg}} = -221.2 \text{ kJ}. \quad \ldots (3)$$

Question 8

Unsteady-state heat transfer problem.

$$I^2 R = mc \frac{\text{d}T}{\text{d}t} + hA_S(T - T_a) \quad \text{for internal-energy generation.}$$

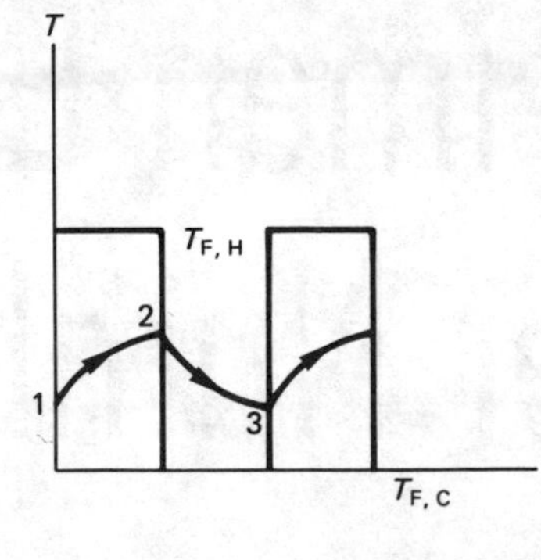

Thus $$\left(\frac{I^2 R}{hA_S} + T_a\right) - T = \frac{mc}{hA_S}\frac{dT}{dt}$$

or $$\frac{T_F - T_2}{T_F - T_1} = e^{-t/t_0} \qquad \text{where } t_0 = \frac{mc}{hA_S}$$

$$t_0 = \frac{\rho \frac{\pi}{4} d^2 lc}{h\pi dl} = \frac{\rho dc}{4h}$$

$$= \frac{19\,000 \frac{kg}{m^3} \times 5 \times 10^{-6} \text{ m} \times 340 \frac{J}{kg\,K}}{4 \times 4000 \frac{W}{m^2\,K}} = 2.019 \text{ milliseconds.}$$

(a) $$\frac{T_2 - T_1}{T_{F,H} - T_1} = 1 - e^{-t/t_0} = \frac{T_3 - T_2}{T_{F,C} - T_2}$$

Also $$\frac{T_2 - T_1}{T_3 - T_2} = -1 \text{ or } -(T_{F,H} - T_1) = (T_{F,C} - T_2).$$

Thus $$(T_{F,H} + T_{F,C}) = (T_1 + T_2) = T_m - 1 + T_m + 1 = 2T_m = 600\ °C.$$

$$T_{F,H} = 600 - 20 = 580\ °C. \quad \text{(a)}$$

(b) $$\frac{580 - 301}{580 - 299} = e^{-t/t_0} \quad \text{or} \quad \frac{t}{t_0} = 0.007\,14.$$

Thus $$t_{ON} = 0.007\,14 \times 2.019 = 0.0144 \text{ millisecond.}$$

$$f = 34\,722 \text{ Hz.} \quad \text{(b)}$$

(c) $$\frac{I^2 R}{hS} = 580\ °C \quad \text{and} \quad R = \frac{\rho l}{\frac{\pi d^2}{4}} = \frac{50 \times 10^{-8}\ \Omega \text{ m} \times 3 \times 10^{-3} \text{ m}}{\frac{\pi}{4}(5 \times 10^{-6})^2 \text{ m}^2}$$

(ρ = resistivity)

or $$R = 76.4\ \Omega.$$

Thus $$V^2 = 76.4\ \Omega \times 580\ °C \times 4000 \frac{W}{m^2\,K} \times \pi \times 5 \times 10^{-6} \text{ m} \times 3 \times 10^{-3} \text{ m} \left[\frac{V^2}{W\,\Omega}\right]$$

$$V^2 = 8.353 \text{ V}^2$$

$$V = 2.89 \text{ V.} \quad \text{(c)}$$

14 Specimen Examination Paper Number 2

14.1 Questions

1 A steam turbine is supplied with steam at the rate of 2000 kg/h. The condition of the steam is 5 bar and 200 °C. The velocity of the steam entering the turbine is negligible. Expansion of the steam within the turbine is adiabatic and the steam leaves with a velocity of 20 m/s. The condition of the steam at exit from the turbine is 0.1 bar and 0.98 dry.

After leaving the turbine the steam enters a condenser where it is condensed and leaves as saturated liquid at 0.1 bar. There is no heat transfer between the condenser and the atmosphere.

Calculate:

(a) the power developed by the turbine,
(b) the heat transfer rate from the condensing steam,
(c) the minimum rate of flow of cooling water to the condenser if its temperature rise is not to exceed 20 K.

Take the specific heat capacity of water to be 4.18 kJ/(kg K).

2 A fuel having a gravimetric analysis carbon of 84 per cent hydrogen 16 per cent is burned with 20 per cent excess air. What is the volumetric analysis of the dry products of combustion, assuming that there is no carbon monoxide?

If the air supply was reduced to the stoichiometric requirement, what proportion of the carbon would burn to carbon dioxide if the dry products of combustion contained 10 per cent carbon monoxide by volume?

3 It is required to reduce the rate of heat transfer through the plasterboard ceiling of a bedroom by the use of a layer of fibreglass insulation material on the upper surface of the plasterboard. The area of the ceiling is 10 m^2, the plasterboard is 12 mm thick, its thermal conductivity is 0.08 W/(m K) and the surface heat transfer coefficient for both upper and lower surfaces is 3.25 W/(m^2 K).

Calculate the rate of heat transfer through the uninsulated ceiling when the temperature difference between the air above and the air below the ceiling is 20 K.

What thickness of fibreglass must be added to reduce the rate of heat transfer to one quarter of that through the uninsulated plasterboard?

The thermal conductivity of the fibreglass is 0.0433 W/(m K) and the surface heat transfer coefficient is 1.05 W/(m^2 K).

4 (a) Write down expressions for the change of internal energy and of enthalpy for a perfect gas. Using these, together with the definition of the property enthalpy, show that for a perfect gas:

$$c_p - c_v = R.$$

(b) For a perfect gas undergoing a reversible, polytropic process obeying the 'law' pv^n = constant show that the initial (p_1, v_1, T_1) and the final (p_2, v_2, T_2) states may be related by the expressions:

$$\frac{T_2}{T_1} = \left(\frac{p_2}{p_1}\right)^{(n-1)/n} ; \quad \frac{T_2}{T_1} = \left(\frac{v_1}{v_2}\right)^{n-1}$$

(c) A perfect gas expands reversibly and adiabatically through a pressure of 1/5. The initial absolute temperature is 1.5 times the final. If the gas constant R = 0.3 kJ/(kg K) detemine c_p and γ.

5 Steam enters a pipe at 4 bar, 0.94 dry at the rate of 4 kg/s. The pipe is 100 m long with an inside diameter of 0.2 m, a wall thickness of 10 mm and a thermal conductivity of 50 W/(m K).

The pipe is lagged with insulating material having a conductivity of 0.9 W/(m K) to a thickness of 50 mm.

Assuming that the inside surface of the pipe is at the temperature of the steam and that the heat transfer coefficient between the lagging surface and the atmosphere is 35 W/(m^2 K), calculate the dryness fraction of the steam at exit from the pipe when the atmospheric temperature is 15 °C. Ignore pipe friction effects.

6 Air flows steadily through a circular duct of 400 mm diameter and at a certain section (1) its velocity, pressure and temperature are respectively 75 m/s, 2 bar and 80 °C. At a second section (2) downstream from section (1) its pressure is 1.85 bar.

Assuming isothermal flow with friction between the two sections, determine:

(a) the mass flow rate of air,
(b) the air velocity at section (2),
(c) the heat transfer between sections (1) and (2),
(d) the frictional drag of the duct walls between sections (1) and (2).

State the assumptions implicit in any expression used.

7 A mass of 1.6 kg of steam contained in a cylinder is initially at 6 bar and 225 °C. If the steam then undergoes a reversible polytropic expansion process according to the law $pv^{1.25}$ = constant until its final pressure is 1 bar, determine:

(a) the final temperature,
(b) the change in internal energy,
(c) the work transfer,
(d) the heat transfer.

Show the process on a sketch of the pressure–volume field.

8 The specific entropy of a fluid changes according to the equation

$$ds = \frac{\delta q_{\text{Rev}}}{T}.$$

Show that for a perfect gas the change in specific entropy between states 1 and 2 is given by:

$$s_2 - s_1 = c_p \ln \frac{T_2}{T_1} - R \ln \frac{p_2}{p_1}.$$

Two vessels A and B each of volume 2.5 m^3 may be connected together by a tube of negligible volume. Vessel A contains air at 6 bar and 90 °C, while B contains air at 3 bar and 200 °C.

Find the final mixture temperature and the change in entropy when vessel A is connected to vessel B, assuming the mixing to be complete and adiabatic.

14.2 Suggested Solutions

Question 1

Steam turbine plant and use of the steady-flow energy equation.

State at turbine entry (1) is superheated. (p. 7 of tables)

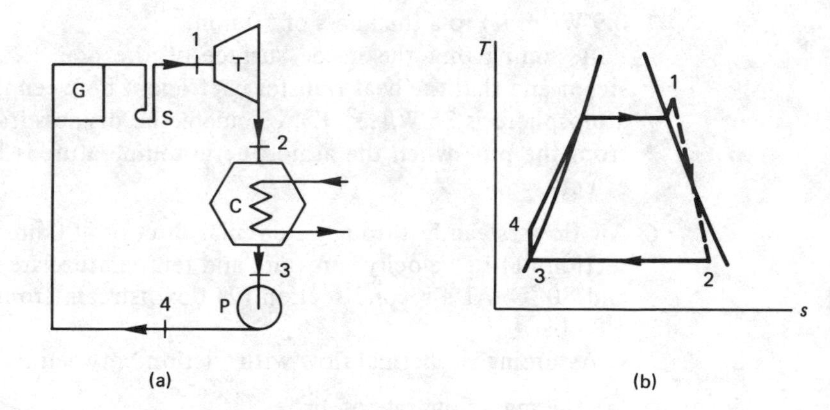

$h_1 = 2857$ kJ/kg (at 5 bar, 200 °C);
$u_1 = 0$ (negligible inlet velocity);
${}_1q_2 = 0$ (adiabatic flow in turbine).

At exit from the turbine (state 2),

$$h_2 = h_{f,2} + x_2 h_{fg,2} = 192 + 0.98\,(2392) = 2536.2\ \frac{\text{kJ}}{\text{kg}} \quad \text{(p. 3 at 0.1 bar, 0.98 dry),}$$

$$h_3 = 192\ \frac{\text{kJ}}{\text{kg}} \quad \text{(condenser exit, saturated liquid at 0.1 bar, p. 3).}$$

Steady-flow energy equation 1 → 2

$${}_1\dot{Q}_2 - {}_1\dot{W}_2 = \dot{m}_S\left[(h_2 - h_1) + \frac{(u_2^2 - u_1^2)}{2} + g\,(z_2 - z_1)\right]$$

which reduces to

$$+\,{}_1\dot{W}_2 = \dot{m}_S\left[(h_1 - h_2) + \frac{u_2^2}{2}\right]$$

or $${}_1\dot{W}_2 = 2000\ \frac{\text{kg}}{\text{h}}\left[\frac{\text{h}}{3600\ \text{s}}\right]\left\{(2857 - 2536.2)\ \frac{\text{kJ}}{\text{kg}} - \frac{1}{2}(20^2)\ \frac{\text{m}^2}{\text{s}^2}\left[\frac{\text{N s}^2}{\text{kg m}}\right]\left[\frac{\text{kJ}}{10^3\ \text{N m}}\right]\right\}$$

$$= 178.1\ \text{kW.} \quad \text{(a)}$$

Steady-flow energy equation 2 → 3

$_2\dot{Q}_3 = \dot{m}_S\,[(h_3 - h_2) + \frac{1}{2}u_2^2]$ assuming $u_3 = 0$. (No work transfer.)

$$= 2000\ \frac{\text{kg}}{\text{h}}\left[\frac{\text{h}}{3600\ \text{s}}\right]\left\{(192 - 2536.2)\ \frac{\text{kJ}}{\text{kg}} - \frac{1}{2}(20^2)\ \frac{\text{m}^2}{\text{s}^2}\left[\frac{\text{N s}^2}{\text{kg m}}\right]\left[\frac{\text{kJ}}{10^3\ \text{N m}}\right]\right\}$$

$= 1302.4$ kW. (b)

$$\dot{m}_w c_{p,w} \Delta T_w = {}_2\dot{Q}_3$$

or

$$\dot{m}_w = \frac{1302.4\ \text{kW}}{4.18\ \frac{\text{kJ}}{\text{kg K}} \times 20\ \text{K}}\left[\frac{\text{kJ}}{\text{s kW}}\right] = 15.58\ \frac{\text{kg}}{\text{s}}. \quad \text{(c)}$$

Question 2

Assume 100 kg fuel containing 84 kg C, 16 kg H_2.

$C + O_2 = CO_2$; $2H_2 + O_2 = 2H_2O$.

12 kg + 32 kg = 44 kg; 4 kg + 32 kg = 36 kg.

O_2 supplied $= 1.2\left[\left(\frac{86}{12} \times 32\right) + \left(\frac{16}{4} \times 32\right)\right]$ kg = 428.76 kg per 100 kg fuel;

N_2 supplied $= \frac{0.7553}{0.2314} \times 428.76 = 1399.5$ kg per 100 kg fuel. (p. 24)

Now stoichiometric O_2 required $= \frac{428.76}{1.2} = 357.3$ kg O_2 per 100 kg fuel.

Thus excess O_2 supplied = 428.76 − 357.3 = 71.46 kg O_2 per 100 kg fuel.

Thus products of combustion are

$\frac{44}{12} \times 86 = 315.33$ kg CO_2 per 100 kg fuel

$+\ \frac{36}{4} \times 16 = 144.00$ kg H_2O per 100 kg fuel ← *ignore* in dry products

+ 71.46 kg O_2 per 100 kg fuel

+ 1399.50 kg N_2 per 100 kg fuel.

Tabulate:

Product	m_i	$m_{w,i}$	$n_i = \frac{m_i}{m_{w,i}}$	$x_i = \frac{n_i}{\Sigma n_i}$	
CO_2	315.33	44	7.167	0.1207	(first answer)
O_2	71.46	32	2.233	0.0376	
N_2	1399.50	28	49.982	0.8417	
			$\Sigma n_i = 59.382$		

If only stoichiometric O_2 is supplied then the dry products will be CO_2, CO and N_2, and for 100 kg fuel there will be

357.3 kg O_2 supplied (see above).

and $\frac{0.7553}{0.2314} \times 357.3 = 1166.2$ kg N_2 supplied.

Let z kg C burn to CO_2 per 100 kg fuel.
Thus $(86 - z)$ kg C burn to CO per 100 kg fuel.

$$2C + \quad O_2 = 2CO$$
$$24 \text{ kg} + 32 \text{ kg} = 58 \text{ kg}.$$

Thus $\left(\frac{86 - z}{24}\right)$ kg C need $\left(\frac{86 - z}{24}\right)$ 32 kg O_2

to produce $\left(\frac{86 - z}{24}\right) \times 56$ kg CO per 100 kg fuel.

And z kg C need $\left(z \times \frac{32}{12}\right)$ kg O_2

to produce $\left(z \times \frac{44}{12}\right)$ kg CO per 100 kg fuel

Tabulate:

Product	m_i	$m_{w,i}$	$n_i = \frac{m_i}{m_{w,i}}$
CO_2	$\frac{44}{12} z$	44	$0.0833z$
CO	$\frac{56}{24}(86 - z)$	28	$7.167 - 0.0833z$
N_2	1166.2	28	41.65
			$\Sigma n_i = 48.817$

$$x_{CO} = 0.1 = \frac{n_{CO}}{\Sigma n_i} = \frac{0.0833z}{48.817}$$

or $z = \frac{48.817 \times 0.1}{0.0833} = 58.6$ kg of C burning to CO per 100 kg fuel.

This proportion of carbon burning to CO only

$$= \frac{58.6}{84} = 0.698 \quad \text{(second answer).}$$

Question 3

Steady-state heat transfer through plane surfaces.

Uninsulated ceiling:

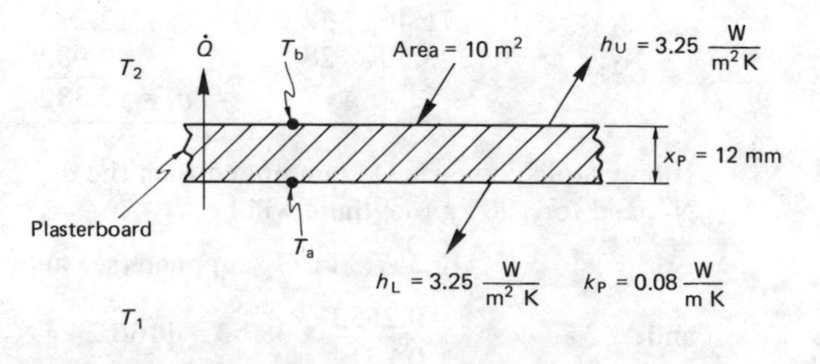

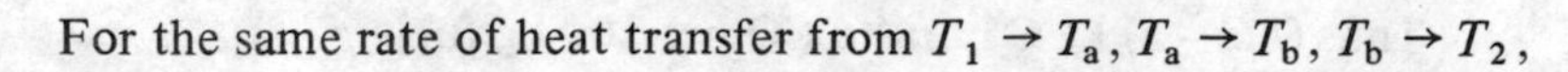

$$T_1 - T_2 = 20 \text{ K given.}$$

For the same rate of heat transfer from $T_1 \rightarrow T_a$, $T_a \rightarrow T_b$, $T_b \rightarrow T_2$,

$$\dot{Q} = h_L A (T_1 - T_a) = \frac{k_P A (T_a - T_b)}{x_P} = h_U A (T_b - T_2).$$

Thus
$$\left.\begin{aligned} T_1 - T_a &= \frac{\dot{Q}}{A\, h_L}, \\ T_a - T_b &= \frac{\dot{Q} x_P}{k_P A}, \\ T_b - T_2 &= \frac{\dot{Q}}{h_U A}. \end{aligned}\right\}$$

Adding:
$$T_1 - T_2 = \dot{Q}\left[\frac{1}{h_L A} \times \frac{x_P}{k_P A} + \frac{1}{h_U A}\right];$$

or
$$\dot{Q} = \frac{T_1 - T_2}{\dfrac{1}{h_L A} + \dfrac{1}{h_U A} + \dfrac{x_P}{k_P A}} = \frac{A(T_1 - T_2)}{\dfrac{1}{h_L} + \dfrac{1}{h_U} + \dfrac{x_P}{k_P}}$$

$$= \frac{10\ \text{m}^2\ (20\ \text{K})}{\left(\dfrac{1}{3.25} + \dfrac{1}{3.25}\right)\dfrac{\text{m}^2\ \text{K}}{\text{W}} + \dfrac{0.012\ \text{m}^2\ \text{K}}{0.08\ \text{W}}} = 261.3\ \text{W}.$$

Now reduced heat transfer $= \dfrac{261.3}{4} = 65.33$ W.

Insulated ceiling:

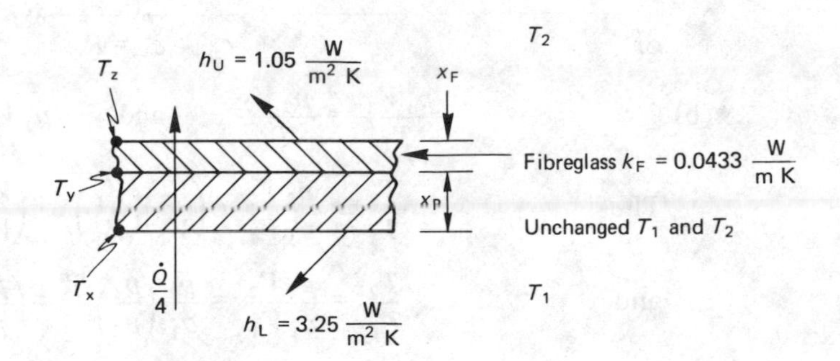

$$\frac{\dot{Q}}{4} = \frac{(T_1 - T_2)}{\dfrac{1}{h_L A} + \dfrac{1}{h_U A} + \dfrac{x_P}{k_P A} + \dfrac{x_F}{k_F A}} \quad \text{where } \dot{Q} = \text{uninsulated value.}$$

Now
$$\left.\begin{aligned} T_1 - T_x &= \frac{\dot{Q}}{h_L A}, \\ T_x - T_y &= \frac{\dot{Q} x_P}{k_P A}, \\ T_z - T_2 &= \frac{\dot{Q}}{h_U A}. \end{aligned}\right\}$$

Adding:
$$T_1 - T_y + T_z - T_2 = \frac{\dot{Q}}{4A}\left[\frac{1}{h} + \frac{1}{h} + \frac{x_P}{k_P}\right];$$

or
$$T_1 - T_y + T_z - T_2 = \frac{65.33\ \text{W}}{10\ \text{m}^2}\left[\left(\frac{1}{3.25} + \frac{1}{1.05}\right)\frac{\text{m}^2\ \text{K}}{\text{W}} + \frac{0.012}{0.08}\ \frac{\text{m}^2\ \text{K}}{\text{W}} = 9.212\ \text{K}.\right]$$

Now $\quad T_1 - T_2 = 20$ K as before.

Thus $\quad T_y - T_z = 20 - 9.212 = 10.788$ K.

Thus $$x_F = \frac{A(T - T_z)\,k_F}{\frac{\dot{Q}}{4}}.$$

$$= \frac{10\ \text{m}^2 \times 10.788\ \text{K} \times 0.0433\ \dfrac{\text{W}}{\text{m K}}}{65.33\ \text{W}} \left[\frac{10^3\ \text{mm}}{\text{m}}\right]$$

$$= 71.5\ \text{mm}.$$

Question 4

(a)

$$E_2 - E_1 = mc_v(T_2 - T_1). \qquad \ldots (1)$$

$$H_2 - H_1 = mc_p(T_2 - T_1). \qquad \ldots (2)$$

Also $$H = E - pV$$

$$H_2 - H_1 = (E_2 - E_1) + (p_2 V_2 - p_1 V_1)$$

or $$H_2 - H_1 = mc_v(T_2 - T_1) + mR(T_2 - T_1). \qquad \ldots (3)$$

Thus, comparing (2) and (3),

$$c_p = c_v + R,$$

or $$c_p - c_v = R. \qquad \ldots (4)$$

(b) $$\frac{p_1 V_1}{T_1} = \frac{p_2 V_2}{T_2} \quad \text{and} \quad p_1 V_1^n = p_2 V_2^n.$$

Thus $$\frac{T_2}{T_1} = \frac{p_2 V_2}{p_1 V_1} = \left(\frac{V_2}{V_1}\right)\left(\frac{V_1}{V_2}\right)^n = \left(\frac{V_1}{V_2}\right)^{n-1} \qquad \ldots (5)$$

and $$\frac{T_2}{T_1} = \frac{p_2 V_2}{p_1 V_1} = \frac{p_2}{p_1}\left(\frac{p_1}{p_2}\right)^{1/n} = \left(\frac{p_2}{p_1}\right)^{(n-1)/n}. \qquad \ldots (6)$$

(c) $$\frac{p_2}{p_1} = \frac{1}{5}; \qquad T_2 = \frac{T_1}{5}; \qquad R = 0.3\ \frac{\text{kJ}}{\text{kg K}};$$

and for a reversible, adiabatic process we can use expression (6), substituting γ for n.

Thus $$\frac{\gamma - 1}{\gamma} \ln \frac{p_2}{p_1} = \ln \frac{T_2}{T_1}$$

or $$\frac{\gamma - 1}{\gamma} = \frac{\ln \dfrac{T_2}{T_1}}{\ln \dfrac{p_2}{p_1}} = \frac{\ln \dfrac{1}{1.5}}{\ln \dfrac{1}{5}} = 0.2516$$

and $$\gamma = \frac{1}{1 - 0.2516} = 1.336. \qquad \ldots (7)$$

$$c_p - c_v = R \text{ or } c_p = \frac{R}{1 - \dfrac{1}{\gamma}} = \frac{0.3\ \dfrac{\text{kJ}}{\text{kg K}}}{0.2516} = 1.192\ \frac{\text{kJ}}{\text{kg K}}. \qquad \ldots (8)$$

Question 5

Radial steady conduction through a cylinder.

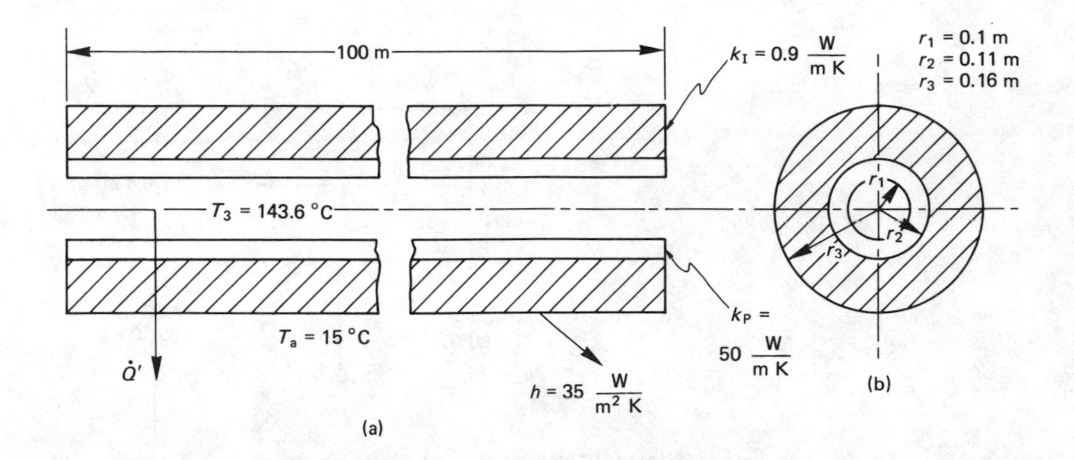

At 4 bar, 0.94 dry, $T_S = 143.6$ °C (p. 4 of tables) $= T_1$ (at r_1) given.

Now
$$T_1 - T_2 = \frac{\dot{Q}' \ln \frac{r_2}{r_1}}{2\pi k_P},$$
$$T_2 - T_3 = \frac{\dot{Q}' \ln \frac{r_3}{r_2}}{2\pi k_I},$$
$$T_3 - T_a = \frac{\dot{Q}'}{\pi d_3 h}.$$

Adding:
$$T_1 - T_a = \dot{Q}' \left[\frac{\ln \frac{r_2}{r_1}}{2\pi k_P} + \frac{\ln \frac{r_3}{r_2}}{2\pi k_I} + \frac{1}{\pi d_3 h} \right]$$

or
$$(143.6 - 15)\ \text{K} = \dot{Q}' \left[\frac{\ln \frac{0.11}{0.1}\ \text{mK}}{2\pi \times 50\ \text{W}} + \frac{\ln \frac{0.16}{0.11}\ \text{mK}}{2\pi \times 0.9\ \text{W}} + \frac{1\ \text{m}^2\ \text{K}}{\pi \times 0.32\ \text{m} \times 35\ \text{W}} \right]$$

or
$$\dot{Q}' = \frac{128.6}{0.000\,303 + 0.0663 + 0.0284} \frac{\text{W}}{\text{m}} = 1353.35 \frac{\text{W}}{\text{m}}.$$

Thus
$$\dot{Q} = 1353.35 \frac{\text{W}}{\text{m}} \times 100\ \text{m} = 135\,335\ \text{W} = 135.33\ \text{kW}.$$

But
$$\dot{Q} = m_S (h_o - h_i)$$
where o ≡ outlet from pipe and i ≡ inlet to pipe (and $\dot{Q}$ is –ve → rejected by steam).

Thus
$$h_o = \frac{\dot{Q}}{m} + h_i = \frac{-135.33\ \text{kW}}{4 \frac{\text{kg}}{\text{s}}} + [605 + 0.94\,(2134)] \frac{\text{kJ}}{\text{kg}} = 2577.13 \frac{\text{kJ}}{\text{kg}},$$

and
$$x_o = \frac{h_o - h_f}{h_{fg}} = \frac{2577.13 - 605}{2134} = 0.924.$$

Question 6

Steady flow of a perfect gas involving both the steady-flow energy equation and the momentum equation.

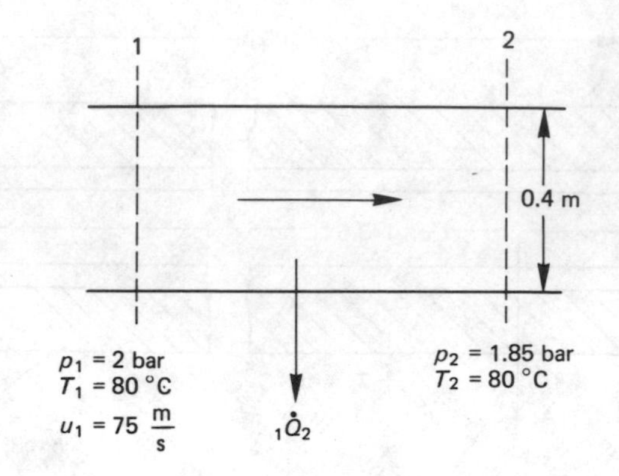

For isothermal flow

$$T_1 = T_2.$$

Work transfer rate = 0.

Gas laws: $\dot{m} = \dfrac{p_1 V_1}{RT_1} = \dfrac{p_1 u_1 A_1}{RT_1}$

$$= \frac{200\ \dfrac{\text{kN}}{\text{m}^2} \times 75\ \dfrac{\text{m}}{\text{s}} \times \pi \times 0.4^2\ \text{m}^2}{0.287\ \dfrac{\text{kJ}}{\text{kg K}} \times 353\ \text{K} \times 4} = 18.61\ \frac{\text{kg}}{\text{s}}. \quad \text{(a)}$$

$$u_2 = \frac{\dot{m}RT_2}{p_2 A_2} = \frac{18.61 \times 0.287 \times 353 \times 4}{185 \times \pi \times 0.4^2} = 81.1\ \frac{\text{m}}{\text{s}}. \quad \text{(b)}$$

Steady-flow energy equation 1 → 2:

$$_1\dot{Q}_2 - {_1\dot{W}_2} = \dot{m}\left[c_p\,(T_2 - T_1) + \frac{(u_2^2 - u_1^2)}{2} + g\,(z_2 - z_1)\right]$$

which reduces to

$$_1\dot{Q}_2 = \dot{m}\ \frac{(u_2^2 - u_1^2)}{2}$$

since flow is isothermal and all else is zero;

i.e. $$_1\dot{Q}_2 = 18.61\ \frac{\text{kg}}{\text{s}} \left(\frac{81.1^2 - 75^2}{2}\right) \frac{\text{m}^2}{\text{s}^2} \left[\frac{\text{N s}^2}{\text{kg m}}\right] \left[\frac{\text{kJ}}{10^3\ \text{N m}}\right] = 8.86\ \text{kW}. \quad \text{(c)}$$

Momentum 1 → 2: $(p_1 \,.\, A_1) + (\dot{m} \,.\, u_1) = (p_2 \,.\, A_2) + (\dot{m} \,.\, u_2) + F$

where F is duct frictional force opposing motion.

$$F = \left(200\ \frac{\text{kN}}{\text{m}^2} - 185\ \frac{\text{kN}}{\text{m}^2}\right) \frac{\pi \times 0.4^2}{4}\ \text{m}^2 + 18.61\ \frac{\text{kg}}{\text{s}}\ (75 - 81.1)\ \frac{\text{m}}{\text{s}} \left[\frac{\text{N s}^2}{\text{kg m}}\right] \left[\frac{\text{kN}}{10^3\ \text{N}}\right]$$

$F = 1.771$ kN. (d)

Question 7

A non-flow example on steam.

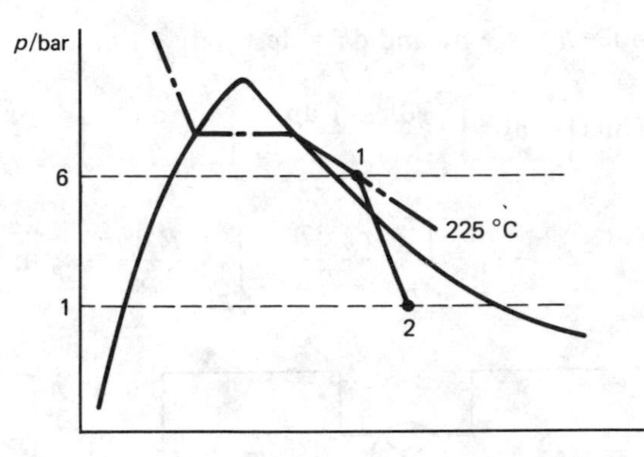

$$v_2 = v_1 \left(\frac{p_1}{p_2}\right)^{1/n} = \left(\frac{0.3522 + 0.3940}{2}\right) \frac{\text{m}^3}{\text{kg}} \, (6)^{1/1.25} = 1.564 \, \frac{\text{m}^3}{\text{kg}}$$

where $v_1 = 0.3731 \, \dfrac{\text{m}^3}{\text{kg}}$ is calculated from p. 7 of tables.

Thus $$x_2 = \frac{v_2}{v_{g,2}} = \frac{1.564}{1.694} = 0.923 \text{ (neglecting } v_f\text{)}.$$

Thus steam is wet at point 2 and $T_2 = T_{sat,2} = 99.6\,°\text{C}$. (a)

$$E_1 = me_1 = 1.6 \text{ kg} \left(\frac{2640 + 2722}{2}\right) \frac{\text{kJ}}{\text{kg}} = 4289.6 \text{ kJ}, \quad \text{(p. 7)}$$

$$E_2 = me_2 = 1.6 \text{ kg} \, [417 + 0.923\,(2506 - 417)] \, \frac{\text{kJ}}{\text{kg}} = 3752.2 \text{ kJ}. \quad \text{(p. 4)}$$

Thus $E_2 - E_1 = -537.3$ kJ. (b)

$${}_1W_2 = \frac{p_1 V_1 - p_2 V_2}{n - 1} = \frac{m\,(p_1 v_1 - p_2 v_2)}{n - 1}$$

$$= \frac{1.6 \text{ kg} \left[\left(600 \, \dfrac{\text{kN}}{\text{m}^2} \times 0.3731 \, \dfrac{\text{m}^3}{\text{kg}}\right) - \left(100 \, \dfrac{\text{kN}}{\text{m}^2} \times 1.564 \, \dfrac{\text{m}^3}{\text{kg}}\right)\right]}{0.25} = 431.7 \text{ kJ}.$$

(c)

Non-flow energy equation:

$${}_1Q_2 = {}_1W_2 + E_2 - E_1 = -105.6 \text{ kJ}. \quad \text{(d)}$$

Question 8

First and second laws: $\mathrm{d}s = \dfrac{\delta q_{\mathrm{R}}}{T} = \dfrac{\mathrm{d}e + p\,\mathrm{d}v}{T} = \dfrac{\mathrm{d}h - v\,\mathrm{d}p}{T}$

since $h = e + pv$ and $\mathrm{d}h = \mathrm{d}e + p\,\mathrm{d}v + v\,\mathrm{d}p$.

$$\text{Thus} \int_{s_1}^{s_2} \mathrm{d}s = \int_1^2 \frac{\mathrm{d}h - v\,\mathrm{d}p}{T} = \int_{T_1}^{T_2} \frac{\mathrm{d}h}{T} - \int_{p_1}^{p_2} \frac{v\,\mathrm{d}p}{T},$$

$$s_2 - s_1 = \int_{T_1}^{T_2} \frac{c_p\,\mathrm{d}T}{T} - \int_{p_1}^{p_2} \frac{R\,\mathrm{d}p}{p} = c_p \ln \frac{T_2}{T_1} - R \ln \frac{p_2}{p_1}.$$

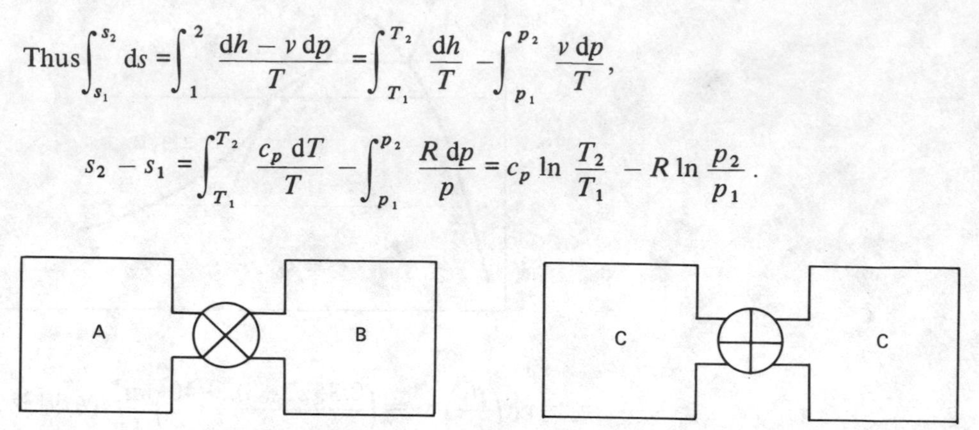

(a) Before (b) After

Energy equation: $E_{\mathrm{A}} + E_{\mathrm{B}} = E_{\mathrm{C}}$ (all else is zero).

$$m_{\mathrm{A}} c_v T_{\mathrm{A}} + m c_v T_{\mathrm{B}} = (m_{\mathrm{A}} + m_{\mathrm{B}}) c_v T_{\mathrm{C}}$$

$$m_{\mathrm{A}} = \frac{p_{\mathrm{A}} V_{\mathrm{A}}}{R T_{\mathrm{A}}} = \frac{600 \dfrac{\mathrm{kN}}{\mathrm{m}^2} \times 2.5\ \mathrm{m}^3}{0.287 \dfrac{\mathrm{kJ}}{\mathrm{kg\ K}} \times 363\ \mathrm{K}} = 14.398\ \mathrm{kg}$$

$$m_{\mathrm{B}} = \frac{p_{\mathrm{B}} V_{\mathrm{B}}}{R T_{\mathrm{B}}} = \frac{300 \times 2.5}{0.287 \times 473} = 5.525\ \mathrm{kg}$$

$$m_{\mathrm{A}} + m_{\mathrm{B}} = 19.903\ \mathrm{kg}$$

$$T_{\mathrm{C}} = \frac{m_{\mathrm{A}} T_{\mathrm{A}} + m_{\mathrm{B}} T_{\mathrm{B}}}{m_{\mathrm{A}} + m_{\mathrm{B}}} = \frac{(14.398 \times 363) + (5.525 \times 473)}{19.903}$$

$$T_{\mathrm{C}} = 393.9\ \mathrm{K}.$$

$$s_{\mathrm{C}} - s_{\mathrm{A}} = c_p \ln \frac{T_{\mathrm{C}}}{T_{\mathrm{A}}} - R \ln \frac{p_{\mathrm{C}}}{p_{\mathrm{A}}}$$

$$= 1.005 \frac{\mathrm{kJ}}{\mathrm{kg\ K}} \ln \frac{393.9}{363} - 0.287 \frac{\mathrm{kJ}}{\mathrm{kg\ K}} \ln \frac{4.5}{6} = 0.1647 \frac{\mathrm{kJ}}{\mathrm{K\ kg}}$$

and $s_{\mathrm{C}} - s_{\mathrm{B}} = 1.005 \ln \dfrac{393.9}{473} - 0.287 \ln \dfrac{4.5}{3} = -0.3003 \dfrac{\mathrm{kJ}}{\mathrm{K\ kg}}$.

Thus $\Delta S = m_{\mathrm{A}} (s_{\mathrm{C}} - s_{\mathrm{A}}) + m_{\mathrm{B}} (s_{\mathrm{C}} - s_{\mathrm{B}})$

$$\Delta S = 14.398\,(0.1647) + 5.525\,(-0.3003) \frac{\mathrm{kJ}}{\mathrm{K}} = 0.7122 \frac{\mathrm{kJ}}{\mathrm{K}}.$$

Recommended Reading

The market is very well supplied with textbooks on engineering thermodynamics and the choice of any one book is very much a subjective one depending upon the student's likes and dislikes.

I have never been able to recommend any of the American books because for one reason or another they do not accord with my own individual teaching practice. This, however, is in no way intended as a reflection on their rigour or excellence.

All in all, one of the most popular textbooks for the whole of a degree course in thermodynamics and heat transfer is *Engineering Thermodynamics, Work and Heat Transfer* by G. F. C. Rogers and Y. R. Mayhew published by Blackwells of Oxford.

A second possibility is the text *Applied Thermodynamics for Engineering Technologists* by T. D. Eastop and A. McConkey, published by Longmans.

Of the two I prefer the former for logical layout and for the very good standard of diagrams in the text. However, both volumes are rigorous and easy to read and cover the material likely to be required in any modern degree course in the subject.

Index